LONDON MATHEMATICAL SOCIETY LECTURE NOTE SERIES

Managing Editor: Professor J.W.S. Cassels, Department of Pure Mathematics and Mathematical Statistics, University of Cambridge, 16 Mill Lane, Cambridge CB2 1SB, England

The titles below are available from booksellers, or, in case of difficulty, from Cambridge University Press.

46   p-adic analysis: a short course on recent work, N. KOBLITZ
50   Commutator calculus and groups of homotopy classes, H.J. BAUES
59   Applicable differential geometry, M. CRAMPIN & F.A.E. PIRANI
66   Several complex variables and complex manifolds II, M.J. FIELD
69   Representation theory, I.M. GELFAND et al
86   Topological topics, I.M. JAMES (ed)
87   Surveys in set theory, A.R.D. MATHIAS (ed)
88   FPF ring theory, C. FAITH & S. PAGE
89   An F-space sampler, N.J. KALTON, N.T. PECK & J.W. ROBERTS
90   Polytopes and symmetry, S.A. ROBERTSON
92   Representation of rings over skew fields, A.H. SCHOFIELD
93   Aspects of topology, I.M. JAMES & E.H. KRONHEIMER (eds)
94   Representations of general linear groups, G.D. JAMES
96   Diophantine equations over function fields, R.C. MASON
97   Varieties of constructive mathematics, D.S. BRIDGES & F. RICHMAN
98   Localization in Noetherian rings, A.V. JATEGAONKAR
99   Methods of differential geometry in algebraic topology, M. KAROUBI & C. LERUSTE
100 Stopping time techniques for analysts and probabilists, L. EGGHE
104 Elliptic structures on 3-manifolds, C.B. THOMAS
105 A local spectral theory for closed operators, I. ERDELYI & WANG SHENGWANG
107 Compactification of Siegel moduli schemes, C.-L. CHAI
109 Diophantine analysis, J. LOXTON & A. VAN DER POORTEN (eds)
110 An introduction to surreal numbers, H. GONSHOR
113 Lectures on the asymptotic theory of ideals, D. REES
114 Lectures on Bochner-Riesz means, K.M. DAVIS & Y.-C. CHANG
116 Representations of algebras, P.J. WEBB (ed)
118 Skew linear groups, M. SHIRVANI & B. WEHRFRITZ
119 Triangulated categories in the representation theory of finite-dimensional algebras, D. HAPPEL
121 Proceedings of Groups - St Andrews 1985, E. ROBERTSON & C. CAMPBELL (eds)
122 Non-classical continuum mechanics, R.J. KNOPS & A.A. LACEY (eds)
128 Descriptive set theory and the structure of sets of uniqueness, A.S. KECHRIS & A. LOUVEAU
129 The subgroup structure of the finite classical groups, P.B. KLEIDMAN & M.W. LIEBECK
130 Model theory and modules, M. PREST
131 Algebraic, extremal & metric combinatorics, M.-M. DEZA, P. FRANKL & I.G. ROSENBERG (eds)
132 Whitehead groups of finite groups, ROBERT OLIVER
133 Linear algebraic monoids, MOHAN S. PUTCHA
134 Number theory and dynamical systems, M. DODSON & J. VICKERS (eds)
135 Operator algebras and applications, 1, D. EVANS & M. TAKESAKI (eds)
137 Analysis at Urbana, I, E. BERKSON, T. PECK, & J. UHL (eds)
138 Analysis at Urbana, II, E. BERKSON, T. PECK, & J. UHL (eds)
139 Advances in homotopy theory, S. SALAMON, B. STEER & W. SUTHERLAND (eds)
140 Geometric aspects of Banach spaces, E.M. PEINADOR & A. RODES (eds)
141 Surveys in combinatorics 1989, J. SIEMONS (ed)
144 Introduction to uniform spaces, I.M. JAMES
145 Homological questions in local algebra, JAN R. STROOKER
146 Cohen-Macaulay modules over Cohen-Macaulay rings, Y. YOSHINO
148 Helices and vector bundles, A.N. RUDAKOV et al
149 Solitons, nonlinear evolution equations and inverse scattering, M. ABLOWITZ & P. CLARKSON
150 Geometry of low-dimensional manifolds 1, S. DONALDSON & C.B. THOMAS (eds)
151 Geometry of low-dimensional manifolds 2, S. DONALDSON & C.B. THOMAS (eds)
152 Oligomorphic permutation groups, P. CAMERON
153 L-functions and arithmetic, J. COATES & M.J. TAYLOR (eds)
155 Classification theories of polarized varieties, TAKAO FUJITA
156 Twistors in mathematics and physics, T.N. BAILEY & R.J. BASTON (eds)
158 Geometry of Banach spaces, P.F.X. MÜLLER & W. SCHACHERMAYER (eds)
159 Groups St Andrews 1989 volume 1, C.M. CAMPBELL & E.F. ROBERTSON (eds)
160 Groups St Andrews 1989 volume 2, C.M. CAMPBELL & E.F. ROBERTSON (eds)
161 Lectures on block theory, BURKHARD KÜLSHAMMER
162 Harmonic analysis and representation theory, A. FIGA-TALAMANCA & C. NEBBIA
163 Topics in varieties of group representations, S.M. VOVSI
164 Quasi-symmetric designs, M.S. SHRIKANDE & S.S. SANE
166 Surveys in combinatorics, 1991, A.D. KEEDWELL (ed)
168 Representations of algebras, H. TACHIKAWA & S. BRENNER (eds)

169	Boolean function complexity, M.S. PATERSON (ed)
170	Manifolds with singularities and the Adams-Novikov spectral sequence, B. BOTVINNIK
171	Squares, A.R. RAJWADE
172	Algebraic varieties, GEORGE R. KEMPF
173	Discrete groups and geometry, W.J. HARVEY & C. MACLACHLAN (eds)
174	Lectures on mechanics, J.E. MARSDEN
175	Adams memorial symposium on algebraic topology 1, N. RAY & G. WALKER (eds)
176	Adams memorial symposium on algebraic topology 2, N. RAY & G. WALKER (eds)
177	Applications of categories in computer science, M. FOURMAN, P. JOHNSTONE & A. PITTS (eds)
178	Lower K- and L-theory, A. RANICKI
179	Complex projective geometry, G. ELLINGSRUD et al
180	Lectures on ergodic theory and Pesin theory on compact manifolds, M. POLLICOTT
181	Geometric group theory I, G.A. NIBLO & M.A. ROLLER (eds)
182	Geometric group theory II, G.A. NIBLO & M.A. ROLLER (eds)
183	Shintani zeta functions, A. YUKIE
184	Arithmetical functions, W. SCHWARZ & J. SPILKER
185	Representations of solvable groups, O. MANZ & T.R. WOLF
186	Complexity: knots, colourings and counting, D.J.A. WELSH
187	Surveys in combinatorics, 1993, K. WALKER (ed)
188	Local analysis for the odd order theorem, H. BENDER & G. GLAUBERMAN
189	Locally presentable and accessible categories, J. ADAMEK & J. ROSICKY
190	Polynomial invariants of finite groups, D.J. BENSON
191	Finite geometry and combinatorics, F. DE CLERCK et al
192	Symplectic geometry, D. SALAMON (ed)
193	Computer algebra and differential equations, E. TOURNIER (ed)
194	Independent random variables and rearrangement invariant spaces, M. BRAVERMAN
195	Arithmetic of blowup algebras, WOLMER VASCONCELOS
196	Microlocal analysis for differential operators, A. GRIGIS & J. SJÖSTRAND
197	Two-dimensional homotopy and combinatorial group theory, C. HOG-ANGELONI, W. METZLER & A.J. SIERADSKI (eds)
198	The algebraic characterization of geometric 4-manifolds, J.A. HILLMAN
199	Invariant potential theory in the unit ball of C^n, MANFRED STOLL
200	The Grothendieck theory of dessins d'enfant, L. SCHNEPS (ed)
201	Singularities, JEAN-PAUL BRASSELET (ed)
202	The technique of pseudodifferential operators, H.O. CORDES
203	Hochschild cohomology of von Neumann algebras, A. SINCLAIR & R. SMITH
204	Combinatorial and geometric group theory, A.J. DUNCAN, N.D. GILBERT & J. HOWIE (eds)
205	Ergodic theory and its connections with harmonic analysis, K. PETERSEN & I. SALAMA (eds)
206	An introduction to noncommutative differential geometry and its physical applications, J. MADORE
207	Groups of Lie type and their geometries, W.M. KANTOR & L. DI MARTINO (eds)
208	Vector bundles in algebraic geometry, N.J. HITCHIN, P. NEWSTEAD & W.M. OXBURY (eds)
209	Arithmetic of diagonal hypersurfaces over finite fields, F.Q. GOUVÊA & N. YUI
210	Hilbert C*-modules, E.C. LANCE
211	Groups 93 Galway / St Andrews I, C.M. CAMPBELL et al
212	Groups 93 Galway / St Andrews II, C.M. CAMPBELL et al
214	Generalised Euler-Jacobi inversion formula and asymptotics beyond all orders, V. KOWALENKO, N.E. FRANKEL, M.L. GLASSER & T. TAUCHER
215	Number theory 1992–93, S. DAVID (ed)
216	Stochastic partial differential equations, A. ETHERIDGE (ed)
217	Quadratic forms with applications to algebraic geometry and topology, A. PFISTER
218	Surveys in combinatorics, 1995, PETER ROWLINSON (ed)
220	Algebraic set theory, A. JOYAL & I. MOERDIJK
221	Harmonic approximation, S.J. GARDINER
222	Advances in linear logic, J.-Y. GIRARD, Y. LAFONT & L. REGNIER (eds)
223	Analytic semigroups and semilinear initial boundary value problems, KAZUAKI TAIRA
224	Computability, enumerability, unsolvability, S.B. COOPER, T.A. SLAMAN & S.S. WAINER (eds)
225	A mathematical introduction to string theory, S. ALBEVERIO, J. JOST, S. PAYCHA, S. SCARLATTI
226	Novikov conjectures, index theorems and rigidity I, S. FERRY, A. RANICKI & J. ROSENBERG (eds)
227	Novikov conjectures, index theorems and rigidity II, S. FERRY, A. RANICKI & J. ROSENBERG (eds)
228	Ergodic theory of Z^d actions, M. POLLICOTT & K. SCHMIDT (eds)
229	Ergodicity for infinite dimensional systems, G. DA PRATO & J. ZABCZYK
230	Prolegomena to a middlebrow arithmetic of curves of genus 2, J.W.S. CASSELS & E.V. FLYNN
231	Semigroup theory and its applications, K.H. HOFMANN & M.W. MISLOVE (eds)
232	The descriptive set theory of Polish group actions, H. BECKER & A.S. KECHRIS
233	Finite fields and applications, S. COHEN & H. NIEDERREITER (eds)
234	Introduction to subfactors, V. JONES & V.S. SUNDER
235	Number theory 1993–94, S. DAVID (ed)
236	The James forest, H. FETTER & B. GAMBOA DE BUEN
237	Sieve methods, exponential sums, and their applications in number theory, G.R.H. GREAVES, G. HARMAN & M.N. HUXLEY (eds)
238	Representation theory and algebraic geometry, A. MARTSINKOVSKY & G. TODOROV (eds)
239	Clifford algebras and spinors, P. LOUNESTO
240	Stable groups, FRANK O. WAGNER
242	Geometric Galois actions I, L. SCHNEPS & P. LOCHAK (eds)

London Mathematical Society Lecture Note Series. 225

A Mathematical Introduction to String Theory
Variational problems,
geometric and probabilistic methods

Sergio Albeverio
Ruhr-Universität, Bochum

Jürgen Jost
Max-Planck Institut, Leipzig

Sylvie Paycha
Université Louis Pasteur, Strasbourg

Sergio Scarlatti
Università di Roma, 'Tor Vergata'

CAMBRIDGE
UNIVERSITY PRESS

CAMBRIDGE UNIVERSITY PRESS
Cambridge, New York, Melbourne, Madrid, Cape Town, Singapore,
São Paulo, Delhi, Dubai, Tokyo, Mexico City

Cambridge University Press
The Edinburgh Building, Cambridge CB2 8RU, UK

Published in the United States of America by
Cambridge University Press, New York

www.cambridge.org
Information on this title: www.cambridge.org/9780521556101

© Cambridge University Press 1997

This publication is in copyright. Subject to statutory exception
and to the provisions of relevant collective licensing agreements,
no reproduction of any part may take place without the written
permission of Cambridge University Press.

First published 1997

A catalogue record for this publication is available from the British Library

ISBN 978-0-521-55610-1 Paperback

Cambridge University Press has no responsibility for the persistence or
accuracy of URLs for external or third-party internet websites referred to in
this publication, and does not guarantee that any content on such websites is,
or will remain, accurate or appropriate. Information regarding prices, travel
timetables, and other factual information given in this work is correct at
the time of first printing but Cambridge University Press does not guarantee
the accuracy of such information thereafter.

Contents

	page
I.0 Introduction	1
I.1 The two-dimensional Plateau problem	7
I.2 Topological and metric structures on the space of mappings and metrics	11
Appendix to I.2: ILH-structures	17
I.3 Harmonic maps and global structures	21
I.4 Cauchy–Riemann operators	31
I.5 Zeta-function and heat-kernel determinants of an operator	36
I.6 The Faddeev–Popov procedure	41
I.6.1 The Faddeev–Popov map	41
I.6.2 The Faddeev–Popov determinant: the case $G=H$	44
I.6.3 The Faddeev–Popov determinant: the general case	46
I.7 Determinant bundles	48
I.8 Chern classes of determinant bundles	59
I.9 Gaussian measures and random fields	66
I.10 Functional quantization of the Høegh-Krohn and Liouville models on a compact surface	75
I.11 Small time asymptotics for heat-kernel regularized determinants	85
II.1 Quantization by functional integrals	92
II.2 The Polyakov measure	96
II.3 Formal Lebesgue measures on Hilbert spaces	101
II.4 The Gaussian integration on the space of embeddings	106
II.5 The Faddeev–Popov procedure for bosonic strings	109
II.6 The Polyakov measure in noncritical dimension and the Liouville measure	113
II.7 The Polyakov measure in the critical dimension $d=26$	117
II.8 Correlation functions	122
References	126
Index	133

Preface

This book is intended as an introduction to certain global analytic and probabilistic aspects of string theory. Nowadays string theory is a domain where mathematics and physics meet, and proceed together concerning certain aspects. However, the theory itself is far from being complete, in fact it is suspended between purely heuristic Ansätze with little hope of mathematical justification and very advanced mathematical ideas. Our aim has been to bring together as far as presently possible the differential-geometric aspects (related to theory of harmonic maps, infinite dimensional differential geometry, Riemann surfaces) and the measure theoretical and probabilistic aspects one encounters when trying to give a sense to the heuristic "Feynman path integrals", so often used not only by physicists but also by mathematicians "to get started".

One of us (J. Jost) worked out a theory of strings with boundary as a quantization of Plateau's problem for minimal surfaces and lectured at several conferences on the geometric aspects of the theory. Two of us, S. Paycha and S. Scarlatti, have been working on relating these aspects with probabilistic ones, in connection with Ph.D. theses in Bochum/Paris and Rome respectively, under the direction of S. Albeverio [Pa1], [Sc]. The probabilistic aspects are connected with the study of a mass zero Høegh-Krohn model, and the first basic study of these aspects was undertaken by S. Albeverio, S. Paycha and S. Scarlatti in collaboration with the late R. Høegh-Krohn.

It was then natural to join efforts and to produce a book which unifies the approaches. We hope our endeavour will be appreciated by the reader. We stress once more that the book presents only a small portion of all aspects of string theory – but we have strived to present this portion as much as possible as a coherent mathematical theory.

Bochum, December 1994

Dedication

This book is dedicated to the dear memory of Raphael Høegh-Krohn (1938–1988). He was a great mathematician and a natural philosopher who on so many occasions was able to show us the correct way, foreseeing so many of the new developments.

I.0 Introduction

In recent years string theory has attracted great interest in physics and mathematics and has become one of the main sources of mutual stimulation and cross fertilization in these areas. Many reasons for this fact can be mentioned:
(a) Classical string theory is concerned with the propagation of classical 1-dimensional curves, "strings", open or closed, in some ambient space, usually $I\!R^d$, under a dynamics which is relativistic and given by a variational principle; see e.g. [GGRT], [N], [Go], [Gr].
This gives interesting connections with the classical calculus of variations, the theory of minimal surfaces (Plateau problem) and the theory of harmonic maps, as developed e.g. in [J2] and [JS].

To explain the basic idea, let us first consider a relativistic point particle of mass m moving freely (i.e. without any acting forces) in 4-dimensional Minkowski space time $M := M^4$. Its trajectory γ_c is a critical point of the action functional

$$A(\gamma) = -m \int_S ds(\gamma) \tag{0.1}$$

where ds is the line element, $m > 0$ is a constant and S is the parameter space (e.g. $S = [0, t]$).

Observing that $A(\gamma)$ is invariant under reparametrizations, it is natural to seek by analogy a classical dynamics for a relativistic string moving in M by looking for critical points of an action functional of the form

$$A(X) = -C \int_S \sqrt{\det \gamma_X}\, d\eta \tag{0.2}$$

where S is the 2-dimensional parameter space for the string, taken to be a 2-dimensional surface embedded in M through $X : S \mapsto M$, $\mu \mapsto X^\mu(\eta)$ $\mu = 1, \ldots, 4$, $\sqrt{\det \gamma_X}\, d\eta$ being the infinitesimal area element of the string as embedded in M, and X being the embedding map from S into M. $\det \gamma_X$ is the determinant of the matrix

$$\frac{\partial X^\mu}{\partial \eta^\alpha} \frac{\partial X^\nu}{\partial \eta^\beta} \mu_{\mu\nu},$$

$\mu_{\mu\nu}$, $\mu, \nu = 1, \ldots, 4$ being the Minkowski metric on M (with signature $(-1, +1, +1, +1)$), and $\eta^\alpha, \alpha = 1, 2$ a parametrization of S. C is a positive constant.

In fact an extension of great relevance in string theory is the one where M is replaced by a d-dimensional Minkowski space or rather its Euclidean version. Interesting connections have been found with problems in algebraic geometry, the theory of Riemann surfaces and number theory; see e.g. [Ma1,2], [Sm], [MoP].

(b) The quantization of string theory gives rise to problems in different areas, according to the method used. E.g. in the case $d = 26$ (resp. $d = 10$ for "fermionic strings") the representation theory of certain infinite-dimensional Lie algebras, Kac–Moody and Virasoro algebras has been used for quantization. This also yields very interesting connections with 2-dimensional conformal fields (and conformal models of statistical mechanics); see e.g. [Ca], [Ka], [Mi], [AHKMTT]. Another approach to quantization of strings, which also works for $d < 26$ (resp. $d < 10$), has been provided by Polyakov ([P1]), who used a heuristic functional integration method, starting from a heuristic Gibbs-like measure [†]. Since in this monograph we shall basically follow Polyakov's approach, trying to give mathematical meaning to its geometric, functional-analytic and probabilistic aspects, let us describe it shortly at this point.

The basic idea actually goes back to Deser–Zumino and Brink–Di Vecchia–Howe and consists in replacing the relativistic action (0.2) by a relativistic action of the form

$$A(X,g) = -C \int_S \sqrt{\det g}\, g^{\alpha\beta} \partial_\alpha X^\mu \partial_\beta X_\mu \, d\eta \qquad (0.3)$$

with g a pseudo-Riemannian locally Minkowski metric on S. Heuristically the critical points of (0.3), in a suitable variational principle, jointly in g and X, are seen to yield the same dynamics as the one obtained from (0.2); see e.g. [GrSW], [BrH].

Polyakov's idea consists in replacing the relativistic action (0.3) by the corresponding Euclidean action $A_E(X,g)$, defined in the same way as $A(X,g)$ but with d-dimensional Minkowski space M^d replaced by Euclidean space $I\!R^d$, and the pseudo-Riemannian metric g by a Riemannian metric, also denoted by g, on S. Let us remark that a similar substitution of relativistic actions by "Euclidean" actions was actually familiar through the Euclidean approach to quantum field theory (see e.g. [S1], [GJ], [AFHKL]) which had permitted a successful use of methods of statistical mechanics. The recovery of the relativistic quantities is then, with strings as well as with quantum fields, to be undertaken at the end, by a suitable "analytic continuation" (exploiting the fact that the usual Euclidean metric in $I\!R^d$ is obtained from the Minkowski metric of signature $(-1,+1,+1,+1,\ldots+1)$ by "analytic continuation" of the first coordination x_0, the time, to "imaginary time" ix_0).

The critical points of A_E are then harmonic maps, and A_E can be looked upon as the classical action of a Euclidean "σ-model" (describing fields X

[†] In this book we shall use equivalently the adjectives "heuristic" and "formal" for objects which are (or are not yet) defined in rigorous mathematical terms.

I.0: Introduction 3

on S with values in M^d). Polyakov's approach to quantization of strings is by formulating interesting quantities associated with the strings essentially as integrals with respect to a heuristic probability measure of the form

$$Z^{-1} \exp[-A_E(X,g)]d_g[X]d[g], \qquad (0.4)$$

with Z a normalization constant and $d_g[X], d[g]$ heuristic "volume measures" on the spaces of all embeddings $X = (X^\mu, \mu = 1, \ldots, d)$ resp. the space of all Riemannian metrics g.

Polyakov's idea was then to exploit a heuristic (infinite-dimensional) differential geometry to "change integration variables" (X,g) to (X,f,φ,t), with f in the space of diffeomorphisms of S and φ in the space of smooth real functions over S (describing "conformal changes of metrics"), and t in the Teichmüller space of S.

The mathematical description of this change of variables in fact involves a series of mathematical problems which connect with complex analysis and certain parts of infinite-dimensional differential geometry. It is one of the aims of this monograph to present this mathematical description and the tools required to perform it. These involve, besides geometric, complex-analytic and global-analytic methods, the control of certain infinite-dimensional determinants.

After these transformations, following Polyakov, the integration with respect to Polyakov's measure (0.4) is heuristically reduced, for $d \leq 26$, after adding a renormalization term to the action $A_E(X,g)$ (which corresponds to taking an ε-regularization $A_E^\varepsilon(X,g)$ of it, adding a term $\mu_\varepsilon^2 \int_S \sqrt{\det g} d\eta$ and taking the limit $\varepsilon \downarrow 0, \mu_\varepsilon^2 \to \infty$), to an integration with respect to a probability measure, a "Liouville measure", associated with a certain 2-dimensional Euclidean quantum field model (Høegh-Krohn's model with mass zero or Liouville's model) and a finite-dimensional measure (on the moduli space for S). This heuristic reduction is described in [P1] and various subsequent publications, e.g. [A], [AN], [D], [DHP1,2], [DNOP], [F], [Gi], [Hab], [Ja], [MN], [Po1,2], [W]. For $d < 26$, the presence of the Liouville measure is essential.

A mathematical description of this measure has been given in [AHKPS1,2,3], adapting methods used before in the massive Høegh-Krohn's model in [AHK1,3] (see also e.g. [AFHKL]). The application of this well-defined measure to Polyakov's heuristic formulation of string theory reduces the range of the embedding dimension to $d \leq 13$ (resp. $d \leq 19$, when using newer results by Kusuoka [Ku]).

Heuristically for the case $d = 26$, called "critical dimension" in the physical literature, the integration with respect to the heuristic Liouville measure is absent and the whole heuristic Polyakov measure (0.4) reduces to a finite-dimensional measure.

I.0: Introduction

In the present monograph we shall give a mathematical treatment of the cases $d \leq 13$ and $d = 26$, presenting all mathematical instruments which are necessary for this study. We shall discuss the case of open as well as closed strings.

Before we start on a more detailed description of the contents of our work, let us shortly mention some topics which we unfortunately had to leave out of our discussion. We discuss only the case of Riemann surfaces S of a fixed genus: we do not sum over the genus (for heuristic discussions of such sums see e.g. [P3]). Also we only mention shortly amplitudes (see e.g. [BrH], [Kn] for heuristic discussion of these quantities); results concerning the Polyakov path integral over bordered surfaces which was investigated in [A], [Jas3,4], will merely be quoted. We also do not enter the discussion of either string field theory or of fermionic and supersymmetric strings (for these cases see e.g. the discussions in [AGGSIS], [BDVH], [P2], [GrSW]). Finally we do not treat such topics as dimensional reduction, connection with infinite dimensional Lie algebras, light cone quantization, statistical physics, conformal field theory and loop spaces. For these topics we can only refer to the current books or specialized work, e.g. [AHKMTT], [ID], [JM], [Ka], [Le], [Mi], [P3], [PS].

Let us now come back to the description of the content of our work. Chapter I presents all mathematical tools needed for the mathematical description of quantized strings to be achieved in chapter II. In section I.1 we briefly discuss the 2-dimensional variational problem, the Plateau problem, associated with the action functional A_E. In section I.2 we introduce the basic topological and metric structures on the space of mappings and metrics. In section I.3 we discuss harmonic maps and global structures on 2-dimensional differentiable manifolds with boundary. In section I.4 we discuss the Cauchy–Riemann operator on Riemann surfaces.

Section I.5 is devoted to a detailed discussion of zeta-functions and heat-kernel determinants of operators. This discussion is useful for handling various infinite-dimensional determinants which arise in the mathematical implementation of Polyakov's reduction procedure, which is the topic of chapter II of our work. We show in particular that in our case the zeta-function and the heat-kernel definitions of determinants are identical up to some constant.

In section I.6 we give a mathematical description of the so called "Faddeev–Popov-procedure" to "factor out" a certain volume term related to the "gauge group" from heuristic functional integrals constructed with actions having large invariance groups. The mathematical definition we give exploits the techniques of regularized determinants discussed in section I.5.

In section I.7 we show how the notion of regularized determinants for elliptic operators on manifolds arises naturally when equipping determinant bundles with a Hermitian metric. This relates in particular to the work of Quillen

I.0: Introduction

and of [BF], [F1,2], which also influences section I.8, where we discuss Chern classes of determinant bundles.

Sections I.9 and I.10 are concerned with a completely different topic, namely probabilistic tools needed for the description of the "quantized Liouville model" on a compact surface. This model describes a two-dimensional quantum field, which arises naturally in the Polyakov reduction studied in the second part of our work (chapter II). In section I.9 we present the Gaussian measures and associated random fields which are useful in handling the free Markov field on a Riemann surface. In section I.10 the construction of the quantized Liouville model (mass zero Høegh-Krohn model) on a Riemann surface is presented.

The first chapter of the book ends with the description of the small time asymptotics. The first part of the book ends with the description of the small time asymptotics for heat-kernel regularized determinants, such as those which enter the "Polyakov reductions" to be discussed in chapter II. All mathematical tools having been presented in chapter I, chapter II concentrates on the mathematical discussion of quantized strings, relating also to heuristic procedures by Polyakov and others.

In section II.1 we describe in general terms the procedures of quantization by functional integrals. The heuristic Polyakov measure and its connections with mathematical objects discussed in chapter I are presented in section II.2. In section II.3 formal Lebesgue measures on Hilbert spaces are introduced and related in section II.4 to integration theory with respect to Gaussian measures on the space of embeddings. The Faddeev–Popov procedure for bosonic strings is explained in section II.5, as an application of the general discussion of section I.6. The connection between the Polyakov measure in noncritical dimension $d < 26$ and the Liouville model discussed in section I.10 is made in section II.6, whereas section II.7 discusses the Polyakov measure in the critical dimension $d = 26$. We conclude with a short discussion of amplitudes, in section II.8.

Clearly much more remains to be done and we hope that our mathematical presentation of string theory will encourage new research and new cross fertilizations between mathematics and physics.

Acknowledgements

It is a great pleasure to thank Professors A.M. Berthier-Boutet de Monvel, L. Boutet de Monvel, G.F. Dell'Antonio, D. Elworthy, Z. Haba, G. Jona-Lasinio and S. Kusuoka for very useful, most stimulating discussions. We also profited from discussions with D. Bennequin, J. Durhuus, L. Gerritzen, F. Gesztesy, R. Gielerak, F. Herrlich, E. Heyerdahl-Hohler, A. Huckleberry, Z. Jaskólski, Ch. Kessler, W. Kirsch, R. Léandre, X. Li-Jost, J. Marion, K. Oeljeklaus, J. Potthoff, M. Röckner, A. Stoll, D. Testard, B. Torresani, S. Wolpert, T. Wurzbacher and B. Zegarliński. We are very grateful to V. Kolokoltsov for a careful reading of the book and many useful suggestions, and also T. Wurzbacher for critically checking a preliminary manuscript of the second author.
Much of this work grew out of the String Seminar at the Ruhr-Universität Bochum 1986–1992, supported by the research project SFB 237 of the DFG and a Bochum–Strasbourg DFG–CNRS project.
We thank Prof. M. Mebkhout, Prof. V. Georgescu and Prof. Z. Haba, for their kind invitations to the Centre de Physique Théorique, Université d'Aix-Marseille II (Luminy), the Poiana-Brasov School and the University of Wroclaw, respectively. Likewise, research visits at MSRI Berkeley (85/86 special program on Riemann surfaces), Princeton (87/88), and Harvard (89/90) have been useful for developing some of the ideas presented here. We also received stimulation by exposing parts of the results in seminars at the Universities of Oslo, Paris VI, VII, Rome I, II, and Warwick as well as at Conferences in Poiana-Brasov (1987), Karpacz (1988), Marseille (1988), Ascona (1988), Washington (1988), and Oberwolfach (Calculus of Variations, 1988). Sylvie Paycha and Sergio Scarlatti gratefully acknowledge financial support through a DAAD-Stipendium and a CNR-grant, respectively.
Sylvie Paycha also would like to thank I.R.M.A. at the Université Louis Pasteur, Strasbourg, for supporting trips to Bochum during the preparation of the book.
Last but not least, we are very grateful to Mrs. Regina Kirchhoff, Mrs. Anja Lohoff, Mr. Georg Schlitt and Mr. Sven Schriewer for their skillful setting of the manuscript.

I.1 The two-dimensional Plateau problem

Let Γ_1, Γ_2 be smooth closed oriented pairwise disjoint Jordan curves[†]

$$\Gamma_1 \subset \{x^1 = x_0^1\}, \quad \Gamma_2 \subset \{x^1 = x_T^1\},$$

(either of them may be empty). We want to minimize the area $A(\Sigma)$ among surfaces $\Sigma \subset \mathbb{R}^d$ with $\partial \Sigma = \Gamma_1 \cup \Gamma_2$. We therefore consider an oriented differentiable surface S with boundary ∂S with a local parameter $z = (z^1, z^2)$ and an embedding [‡]

$$X : S \to \mathbb{R}^d$$

mapping ∂S monotonically and with preserved orientation onto $\Gamma_1 \cup \Gamma_2$.

Then [§]

$$A(X(S)) = \int_S (\det(\frac{\partial X^\alpha}{\partial z^i} \frac{\partial X^\alpha}{\partial z^j}))^{\frac{1}{2}} dz^1 dz^2. \tag{1.1}$$

We write this as $A(X, S)$. This expresses the fact that the functional A depends on two variables, namely the surface S (or, more precisely, its topological type or genus) and the embedding X. We put

$$\gamma_{ij} := \frac{\partial X^\alpha}{\partial z^i} \frac{\partial X^\alpha}{\partial z^j},$$

$\gamma := (\gamma^{ij}) := (\gamma_{ij})^{-1}$. The Euler–Lagrange equations of (1.1) are

$$\frac{\partial}{\partial z^i} \left((\det \gamma)^{\frac{1}{2}} \gamma^{ij} \frac{\partial X^\alpha}{\partial z^j} \right) = 0, \quad \alpha = 1, \ldots d. \tag{1.2}$$

This equation is nonlinear, because the metric (γ_{ij}) depends on the embedding X; namely it is the metric on $X(S)$ induced by the ambient Euclidean metric of $\mathbb{R}^d$. The nonlinearity of the equations results from the fact that we are trying to determine the embedding X and the metric (γ_{ij}) on S at the same time. Another serious difficulty is that A has the full diffeomorphism group of Σ as invariance group. Namely, if

$$\phi : \Sigma \to \Sigma$$

[†] The requirement that Γ_1 and Γ_2 are contained in fixed hyperplanes will play no role in the sequel and may be abandoned.
[‡] For physical reasons, at the moment, only embeddings will be considered. Later on, however, this requirement will not be maintained.
[§] Here and in the sequel, we use standard summation conventions.

is a diffeomorphism, then

$$A(X \circ \phi, \Sigma) = A(X, \Sigma). \tag{1.3}$$

This has the consequence that one cannot expect much control over a minimizer X, in particular no good a priori estimates.

We therefore consider a different model. We let $(g_{ij})_{i,j=1,2}$ be a metric on S and put, with $(g^{ij}) = (g_{ij})^{-1}$, $\det g = g_{11}g_{22} - g_{12}^2$, for $X : S \to \mathbb{R}^d$

$$D(X, g) := \frac{1}{2} \int_S g^{ij} \frac{\partial X^\alpha}{\partial z^i} \frac{\partial X^\alpha}{\partial z^j} \sqrt{\det g}\, dz^1 dz^2 \tag{1.4}$$

($z = (z^1, z^2)$ is again a local parameter). A solution of the corresponding problem

$$D(X, g) \to \min$$

is then harmonic w.r.t. the metric g:

$$\frac{1}{\partial z^i}\left(g^{ij}\sqrt{\det g}\frac{\partial}{\partial z^j}X^\alpha\right) = 0, \quad \alpha = 1,\ldots d \tag{1.5}$$

and conformal:

$$T_{ij} := \frac{\partial X^\alpha}{\partial z^i}\frac{\partial X^\alpha}{\partial z^j} - \frac{1}{2}g_{ij}g^{kl}\frac{\partial X^\alpha}{\partial z^k}\frac{\partial X^\alpha}{\partial z^l} = 0 \tag{1.6}$$

$((T_{ij})$ is called the energy momentum tensor). We point out that $D(X, g)$ is invariant under

- diffeomorphisms of S; $\phi : S \to S$, a diffeomorphism, transforms g_{ij} into $g_{kl}(\phi(z))\frac{\partial \phi^k}{\partial z^i}\frac{\partial \phi^l}{\partial z^j}$, and invariance of $D(X, g)$ follows from the change of variables formula

- conformal transformations: $g_{ij} \to \lambda(z)g_{ij}$, where $\lambda(z) > 0$; this invariance actually implies

$$g^{ij}T_{ij} = T_i^i = 0,$$

  i.e. the energy momentum tensor is tracefree

- isometries of $\mathbb{R}^d$ (translations and rotations applied to X).

Invariance under diffeomorphism means

$$D(X, g) = D(X \circ \phi, \phi^* g), \tag{1.7}$$

$\phi : S \to S$ a diffeomorphism. If we keep g fixed, however, then D as a function of X is no longer invariant under the action of the diffeomorphism group.

I.1: The two-dimensional Plateau problem

Conformal invariance means that g is determined only up to a conformal factor. Therefore, the problem can also be considered in the following way. Let Σ be a Riemann surface, diffeomorphic to S, with $z = z^1 + iz^2 = u + iv$ now a local <u>conformal</u> parameter (see e.g [B], [FK]) and put

$$D(X, \Sigma) := \frac{1}{2} \int_\Sigma \frac{\partial X^\alpha}{\partial z^i} \frac{\partial X^\alpha}{\partial z^i} dz = \frac{1}{2} \int_\Sigma |\nabla X|^2 dz.$$

D is now only invariant under <u>conformal</u> diffeomorphisms of Σ. We also note $D(X, \Sigma) \geq A(X, \Sigma)$, with equality if and only if $X : \Sigma \to \mathbb{R}^d$ is conformal. As D is quadratic in X and involves no square roots it is mathematically much easier to handle than A.

A maximum X of $D(X, \Sigma)$ for fixed Σ first of all is harmonic, i.e.

$$\Delta X^\alpha = 0, \quad \alpha = 1, \ldots d. \tag{1.8}$$

This implies that ($X_u^2 = X_u^\alpha X_u^\alpha$ etc.)

$$4X_z^2 dz^2 = (X_u^2 - X_v^2 - 2iX_u X_v)dz^2 =: \Phi dz^2 \tag{1.9}$$

is a <u>holomorphic quadratic differential</u> on Σ (see e.g. [L]). Moreover, because we minimize over all $X : \Sigma \to \mathbb{R}^d$ mapping $\partial \Sigma$ monotonically onto $\Gamma_1 \cup \Gamma_2$, Φdz^2 is real on $\partial \Sigma$. This means that if we choose locally $z = u + iv$ in such a way that u is tangential to $\partial \Sigma$ and v normal, then

$$0 = \text{Im } \Phi dz^2_{|\partial \Sigma} = \text{Im } ((X_u^2 - X_v^2 - 2iX_u X_v)du^2), \tag{1.10}$$

i.e.

$$X_u \cdot X_v = 0 \quad \text{on } \partial \Sigma. \tag{1.11}$$

At this point, we have to discuss the regularity question. While it is classical that X as a harmonic function is regular in the interior, it follows from Hildebrandt's theorem and suitable extensions thereof (cf. [J2]) that because of the boundary condition (1.11) X is of class $C^{k,\alpha}$ on $\partial \Sigma$, provided Γ_1 and Γ_2 enjoy this regularity property ($k \in \mathbb{N}$, $0 < \alpha < 1$).

$X(\Sigma)$, however, does not represent a minimal surface, unless X is conformal, i.e.

$$\Phi dz^2 = 0. \tag{1.12}$$

This can only be achieved by also varying Σ. Actually (as seen by a simple computation using isothermal coordinates)

$$\frac{d}{d\Sigma} D(X, \Sigma),$$

where X solves (1.8), can be identified with Φdz^2. (Only if Σ were the sphere or the unit disk would the holomorphicity of Φdz^2 and (1.11) automatically imply $\Phi dz^2 = 0$). We thus also have to minimize $D(X, \Sigma)$ over the conformal structure of Σ. The space of these conformal structures on a surface of genus p is the moduli space to be discussed in more detail below. It is not a compact space but can be compactified by surfaces of genus less than p. Therefore, our minimization problem naturally requires us to consider a third parameter, besides X and the conformal structures Σ, namely the genus [†].

The following observation, however, is useful. We can now start with a fixed map
$$X_0 : \partial S \to \Gamma_1 \cup \Gamma_2$$
which is monotonic and respects orientations, and look only at maps
$$\tilde{X} : S \to \mathbb{R}^d$$
with $\tilde{X}_{|\partial S} = X_0$. The Plateau boundary condition is then achieved by the operation of the group D of diffeomorphisms of S. Namely, if we have a pair (Y, g), $Y : S \to \mathbb{R}^d$ as before, g a metric on S, then we can find a diffeomorphism $\phi : S \to S$ with
$$Y \circ \phi_{|\partial S} = X_0$$
and look at the pair
$$(Y \circ \phi, \ \phi^* g).$$
Therefore, the (nonlinear) Plateau boundary problem is reduced to a (linear) Dirichlet problem plus the action of the diffeomorphism group.

For the sequel, it will be essential not to let X_0 be completely arbitrary, but to let it be harmonic in the interior of S w.r.t. a given metric g. Then we can write $\tilde{X}$ as above as
$$\tilde{X} = X_0 + X, \text{ with } X_{|\partial S} = 0,$$
and since X_0 is harmonic, using integration by parts we obtain
$$D(\tilde{X}, g) = D(X_0, g) + D(X, g).$$

[†] Actually, the requirement that X is embedded makes it necessary to increase the genus as well but this will not be discussed here.

I.2 Topological and metric structures on the space of mappings and metrics

Let S be a compact oriented 2-dimensional differentiable manifold with boundary ∂S diffeomorphic to $\Gamma_1 \cup \Gamma_2$. We shall require below that all metrics are bounded in the appropriate norm on S, including ∂S. For the moment, we fix a metric g_0 on S. We also fix some embedding

$$X_0 : S \to \mathbb{R}^d$$

mapping ∂S monotonically and with preserved orientation onto $\Gamma_1 \cup \Gamma_2$. We also require that X_0 is harmonic, i.e. $\triangle X_0 = 0$. We then have for an arbitrary $X \in H_0^s(S, \mathbb{R}^d)$, $1 \leq s < \infty$ and a metric g

$$D(X + X_0, g) = D(X, g) + D(X_0, g), \qquad (2.1)$$

since $\int g^{ij} \frac{\partial X_0^\alpha}{\partial z^i} \frac{\partial X^\alpha}{\partial z^j} \sqrt{\det g}\, dz = 0$ because X_0 is harmonic and $X_{|\partial S} = 0$. We therefore consider the affine spaces

$$\mathcal{F}^s := X_0 + H_0^s(S, \mathbb{R}^d);$$

here, for simplicity, we do not require the embedding property any more. Integration being defined w.r.t. the volume form of g_0, $\mathcal{F}^s$ is an affine Hilbert space for $1 \leq s < \infty$. Moreover,

$$\mathcal{F} := \mathcal{F}^\infty$$

is an affine Fréchet space. As will be discussed below in the appendix to this chapter, it also has a strong ILH-structure. The tangent space of $\mathcal{F}^s$ is the corresponding linear space

$$T_X \mathcal{F}^s = H_0^s(S, \mathbb{R}^d), \ X \in \mathcal{F}^s. \qquad (2.2)$$

If instead of g_0 we take any other metric g, satisfying the above mentioned boundedness condition, we obtain equivalent Sobolev structures. Of particular importance will be the Hilbert-structure defined by g: for $V, W \in T_X \mathcal{F}^1$

$$(V, W)_g := \int_S V^\alpha(z) W^\alpha(z) \sqrt{\det g}\, dz. \qquad (2.3)$$

For $s \geq 2$, this product introduces only a weak Riemannian structure on $\mathcal{F}^s$, and $\mathcal{F}^s$ in the corresponding topology is no longer complete.

For our purposes it is not legitimate to fix a reference metric g_0; we have to consider all metrics g instead, and we thus have to study how $(V, W)_g$ changes if g changes. We observe first

Lemma 2.1: $(.,.)_g$ is not invariant under conformal transformations

$$(g_{ij}(z) \to \lambda(z)g_{ij}(z), \lambda > 0).$$

∎

We now turn to the action of

$$\mathcal{D}^l := \{\text{diffeomorphisms } \phi : S \to S \text{ of Sobolev class } H^l\}.$$

Again, Sobolev norms depend on the choice of a metric g_0, but any two metrics lead to equivalent norms. In case $l \leq 2$, $\phi \in \mathcal{D}^l$ need not to be differentiable so that we actually have to take the completion of the space of C^∞-diffeomorphisms with respect to the H^l-norm. The action is given by

$$\mathcal{F}^s \times \mathcal{D}^l \to \mathcal{F}^t, t \leq \min(l,s)$$

$$(X, \phi) \to X \circ \phi. \tag{2.4}$$

This map is of class C^∞ in X, for fixed ϕ, and of class C^{s-t} in ϕ; namely, if one differentiates $X \circ \phi_t$, with respect to t, m times, one obtains m derivatives of X from the chain rule. Usually, we shall put $s = t = 1$, so that the action of $\mathcal{D}^l$ is only continuous. But:

Lemma 2.2: The above weak Riemannian structure on $\mathcal{F}^s$ is invariant under the action of $\mathcal{D}^l$, namely

$$(V, W)_g = (V \circ \phi, W \circ \phi)_{\phi^* g} \quad \text{for } \phi \in \mathcal{D}^l$$

$$\int_S V(\phi(z))W(\phi(z))(\det \phi^* g)^{\frac{1}{2}} dz$$

$$= \int V(\zeta)W(\zeta)(\det g(\zeta))^{\frac{1}{2}} d\zeta \quad \text{with } \zeta = \phi(z),$$

since $(\phi^* g_{ij})(z) = g_{kl}(\phi(z)) \frac{\partial \phi^k}{\partial z^i} \frac{\partial \phi^l}{\partial z^j}$.

∎

We now turn to the space of Riemannian metrics on S:

$$\mathcal{M}^k := \{(g_{ij})_{i,j=1,2} \text{ positive definite symmetric } 2 \times 2 \text{ tensor of class } H^k\}. \tag{2.5}$$

Here, we need to assume $k \geq 2$ as in this case by the Sobolev embedding theorem the metrics are of class L^∞. This is needed in order to uniformize the metrics later on, and also for the definition of $(\cdot,\cdot)_g$, cf. (2.3) above.

I.2: Topological and metric structures on mappings and metrics

Again, we put

$$\mathcal{M} := \mathcal{M}^\infty = \bigcap_{k \geq 2} \mathcal{M}^k, \tag{2.6}$$

the space of smooth Riemannian metrics. The action of the diffeomorphism group is given as follows:

$$\mathcal{M}^k \times \mathcal{D}^l \to \mathcal{M}^m, \, m \leq \min(k, l-1)$$
$$((g_{ij}), \phi) \to (g_{kl}(\phi(z)) \frac{\partial \phi^k}{\partial z^i} \frac{\partial \phi^l}{\partial z^j}) = \phi^*(g_{ij}). \tag{2.7}$$

Again, the action is smooth in g, for fixed ϕ, and of class $k - m$ in ϕ. The corresponding action

$$\mathcal{M} \times \mathcal{D} \to \mathcal{M}$$

is smooth in both variables. Here we have set $\mathcal{D} = \bigcap_{l \geq 2} \mathcal{D}^l$. The tangent space of $\mathcal{M}^k$ at g is given by

$$T_g \mathcal{M}^k = \{(h_{ij})_{i,j=1,2} \text{ symmetric } 2 \times 2 \text{ tensor of class } H^k\}.$$

The metric (g_{ij}) defines products on all tensor spaces over S, and consequently a weak Riemannian metric on $\mathcal{M}^k$ is given by

$$((h_{ij}), (k_{ij}))_g := \int_S g^{ij} g^{lm} h_{il} k_{jm} (\det g)^{\frac{1}{2}} dz \tag{2.8}$$

Analogously to Lemmas 2.1, 2.2, we get:

Lemma 2.3: $((h_{ij}), (k_{ij}))_g$ is invariant under the action of $\mathcal{D}^l$, but not invariant under conformal transformations of g.

∎

We now want to describe certain decompositions of $\mathcal{M}^k$. $h = (h_{ij}) \in T_g \mathcal{M}^k$ can be decomposed into its trace part and a tracefree part:

$$h = \rho g + h', \, \rho : S \to \mathbb{R} \tag{2.9}$$

with

$$h'_{ij} = \frac{1}{2}(\delta_i^k \delta_j^l + \delta_i^l \delta_j^k - g_{ij} g^{kl}) h_{kl}$$
$$=: G_{ij}^{kl} h_{kl}. \tag{2.10}$$

This decomposition is pointwise orthogonal and hence in particular orthogonal with respect to $(.,.)_g$. Actually, this means that a somewhat more general metric on $T_g \mathcal{M}^k$ can be defined for any $\kappa = \text{const.} > 0$ as

$$((h_{ij}), (k_{ij}))_{g,\kappa} := \int (G^{ijlm} + \kappa g^{ij} g^{lm}) h_{ij} k_{lm} (\det g)^{\frac{1}{2}} dz. \tag{2.11}$$

I.2: Topological and metric structures on mappings and metrics

Since

$$G^{ijkl} = g^{im}g^{jn}G^{kl}_{mn} = \frac{1}{2}(g^{ik}g^{jl} + g^{il}g^{jk} - g^{ij}g^{kl})$$

and because of the symmetry of (h_{ij}), (k_{ij}),

$$(.,.)_g = (.,.)_{g,\frac{1}{2}}.$$

As the decomposition (2.9) is orthogonal (with respect to any $(.,.)_{g,\kappa}$), for our subsequent considerations it will suffice to study the case $\kappa = \frac{1}{2}$, because the more general case does not lead to any new phenomena.

By Poincaré's uniformization theorem [†], in the most interesting case where the Schottky double S^d of S (defined as S and a copy of S with opposite orientation identified along their boundaries) has genus $p \geq 2$, every g is conformally equivalent to a unique hyperbolic metric (of curvature -1) for which ∂S consists of closed geodesics. We denote the space of these metrics by

$$\mathcal{M}^k_{-1}.$$

We note that a pure trace tensor ρg_{ij}, $\rho : S \to \mathbb{R}$, is transversal to $\mathcal{M}^k_{-1}$. As above, we have actions

$$\mathcal{M}^k_{-1} \times \mathcal{D}^l \to \mathcal{M}^m_{-1}, \quad \mathcal{M}_{-1} \times \mathcal{D} \to \mathcal{M}_{-1} \qquad (2.12)$$

since the action of $\mathcal{D}^l$ is by isometries and hence preserves the curvature. We want to turn to the infinitesimal version of this action. Let a smooth family $(\phi_t) \subset \mathcal{D}$, $\phi_0 = \text{id}$, be generated by a vector field

$$V(z) = \frac{d}{dt}\phi_t(z)_{|t=0}.$$

We compute

$$\frac{d}{dt}(\phi_t^* g_{ij})_{|t=0} = g_{ij,k}V^k + g_{ik}V^k_{,z^j} + g_{jk}V^k_{,z^i} \qquad (2.13)$$

(where $V^k_{,z^j} = \frac{\partial V^k}{\partial z^j}$ etc.). In invariant notation, this reads

$$\frac{d}{dt}(\phi_t^* g)_{|t=0} = L_V g, \text{ where } L \text{ is the Lie derivative.} \qquad (2.14)$$

Also, ∇ being the covariant derivative,

$$(\frac{d}{dt}\phi_t^* g_{|t=0})_{ij} = g_{il}(\nabla_{\frac{\partial}{\partial z^j}}V)^l + g_{jl}(\nabla_{\frac{\partial}{\partial z^i}}V)^l. \qquad (2.15)$$

[†] see [J1] or [J2] for details when the metric is not C^∞.

I.2: Topological and metric structures on mappings and metrics

We now want to determine those symmetric tensors (h''_{ik}) that are tracefree and orthogonal to all tensors of the form $L_v g$ with respect to (2.13). To facilitate computations, we use isothermal coordinates so that

$$g_{ij}(z) = \rho^2(z)\delta_{ij}, \quad \rho^2(z) > 0. \tag{2.16}$$

The orthogonality condition, valid for all V,

$$0 = \int g^{ij} g^{kl} h''_{ik}(g_{jl,m} V^m + 2g_{jm} V^m_{z^l})(\det g)^{\frac{1}{2}} dz \tag{2.17}$$

(where we have already used the symmetry of (h''_{ik})) then becomes

$$0 = \int \frac{1}{\rho^2} h''_{ik}(g_{ik,m} V^m + 2\rho^2 V^i_{z^k}) dz. \tag{2.18}$$

The first term vanishes because (h''_{ik}) is tracefree and because of (2.16). For the second term, we first assume that V has its support in the interior of S. Integrating by parts, we obtain the condition

$$\frac{\partial}{\partial z^k}(h''_{ik}) = 0, \quad i = 1, 2, \tag{2.19}$$

i.e. h''_{ik} is divergencefree. In arbitrary coordinates, this condition becomes

$$\text{div}_g(g^{jk} h''_{ik}) = \frac{1}{(\det g)^{\frac{1}{2}}} \frac{\partial}{\partial z^j}((\det g)^{\frac{1}{2}} g^{jk} h''_{ik}) = 0, \quad i = 1, 2. \tag{2.20}$$

If we now allow V with arbitrary support, we also get a boundary term from the integration by parts leading to (2.19), and this boundary term then also has to vanish.

Since we require that all diffeomorphisms map ∂S into itself, V has to be tangential to ∂S. We choose our isothermal coordinates $z = x + iy$ near the boundary in such a way that x is tangential and y is normal to ∂S. Thus,

$$V^2 = 0 \text{ on } \partial S \tag{2.21}$$

where $V = V^1 \frac{\partial}{\partial x} + V^2 \frac{\partial}{\partial y}$. The boundary term that occurs from integrating (2.18) by parts is then

$$\int_{\partial S} h''_{12} V^1 dx$$

and since this again has to vanish for all V, we conclude

$$h''_{12} = 0 \text{ on } \partial S. \tag{2.22}$$

We want to interpret these conditions on h''. First of all,

$$h'' = \begin{pmatrix} h''_{11} & h''_{12} \\ h''_{21} & h''_{22} \end{pmatrix} =: \begin{pmatrix} u & v \\ v & -u \end{pmatrix}$$

as h'' is symmetric and tracefree. (2.19) then becomes

$$u_x = -v_y, \quad u_y = v_x, \tag{2.23}$$

where $z = x + iy$ is a conformal parameter. Thus,

$$u - iv$$

is holomorphic, which means that

$$h'' = u dx^2 - u dy^2 + 2v dx dy = \operatorname{Re}\left((u - iv)(dx + idy)^2\right)$$

is the real part of a holomorphic quadratic differential

$$\Phi dz^2 := (u - iv) dz^2.$$

(2.22) becomes, choosing x tangential to ∂S,

$$v = 0 \text{ on } \partial S, \tag{2.24}$$

i.e. Φdz^2 is real on ∂S.

(2.22) and the preceding interpretation of the fact that h'' is divergencefree mean that the tangent directions of $\mathcal{M}^k$ which are normal both to the conformal reparametrizations and the action of the diffeomorphism group are given by real parts of holomorphic quadratic differentials, where holomorphicity is defined with respect to the conformal structure induced by the metric which is the base point of the tangent space we are looking at. On ∂S, these differentials have to be real.

Let now

$$\Phi dz^2 = (u_1 - iv_1) dz^2, \quad \Psi dz^2 = (u_2 - iv_2) dz^2$$

be two such differentials. Their product given by (2.18) is evaluated, again using conformal coordinates as in (2.19), as follows.

$$\begin{aligned}(\Phi dz^2, \Psi dz^2)_g &= 2 \int (u_1 u_2 + v_1 v_2) \cdot \frac{1}{\rho^2(z)} dz d\bar{z} \tag{2.25}\\ &= 2 \operatorname{Re} \int \Phi \bar{\Psi} \frac{1}{\rho^2(z)} dz d\bar{z}\end{aligned}$$

I.3: Harmonic maps and global structures

(with Re denoting the real part). This product is the Weil–Petersson metric on the space of holomorphic quadratic differentials on (S,g). The preceding discussion of $\mathcal{M}^k$ is adapted from [T1,2].

Let us also recall that the infinitesimal generators of diffeomorphisms are tangent vector fields V which are tangential to ∂S. The metric on the space of tangent fields is given by

$$(V,W)_g := \int_S g_{ij} V^i W^j (\det g)^{\frac{1}{2}} dz. \quad (2.26)$$

Again, this is not conformally invariant.

Appendix to I.2: ILH-structures

In [O], Omori proved that the diffeomorphism group $\mathcal{D}$ of a compact manifold M without boundary is a strong ILH-Lie group, a concept to be defined below (for a survey, see also [M]). In this section we present the relevant definitions and sketch an extension of this result to the case where M has a boundary.

Definition A.2.1: Let $\underline{E}$ be a topological linear space. $\underline{E}$ is called an ILH-space,

$$\underline{E} = \varprojlim E_\nu \text{ (inverse limit)},$$

if $\{E_\nu\}_{\nu \in \mathbb{N}}$ is a sequence of Hilbert spaces, with $E_\nu \subset E_\mu$ for $\nu \geq \mu$, for which the inclusion maps

$$i_\nu : E_\nu \to E_{\nu-1}$$

are bounded linear operators.

A topological space $\underline{X}$ is called a C^k-ILH-manifold modeled on $\underline{E}$ provided the following two conditions hold.

(A1) $\underline{X}$ is an inverse limit of Hilbert manifolds X_ν modeled on E_ν with $X_\nu \subset X_\mu$ for $\nu \geq \mu$.

(A2) For each $x \in \underline{X}$ there exist open neighborhoods $U_\nu(x)$ of x in X_ν and homeomorphisms $\Psi_\nu : U_\nu(X) \to V_\nu \subset E_\nu$ yielding C^k-coordinate systems around x in X_ν and satisfying

$$U_{\nu+1}(x) \subset U_\nu(x) \text{ and, for } y \in U_{\nu+1}(x),\ \Psi_{\nu+1}(y) = \Psi_\nu(y).$$

$\underline{X}$ is called a strong C^k-ILH-manifold modeled on $\underline{E}$ if in addition

(A3) $\lim U_\nu(x)$ is an open neighborhood of x in $\underline{X}$.

For a C^k-ILH-manifold $\underline{X}$ ($k \geq 1$), the ILH-tangent bundle of $\underline{X}$ is defined as
$$T\underline{X} = \varprojlim TX_\nu.$$

A map $\underline{\Phi} : \underline{X} \to \underline{Y}$ between C^k-ILH-manifolds is called C^k-ILH-differentiable if for every ν there exist $\mu(\nu)$ and a C^k-map $\Phi_\nu : X_{\mu(\nu)} \to Y_\nu$ with
$$\Phi_{\nu+1}(x) = \Phi_\nu(x) \text{ for every } x \in X_{\mu(\nu+1)}$$
and
$$\underline{\Phi} = \varprojlim \Phi_\nu,$$
i.e. $\underline{\Phi}$ is the inverse limit of C^k-differentiable maps. Finally, we obtain C^∞-ILH-objects by requiring that the preceeding conditions are satisfied for all $k \geq 0$.

A topological group is called (strong) ILH-Lie group if it is a C^∞- (strong-) ILH-manifold and the group operations are C^∞-ILH-mappings.

For $x \in \underline{X}$, we put
$$T_x\underline{X} := \varprojlim T_x X_\nu$$
and for $\underline{\Phi} : \underline{X} \to \underline{Y}$ C^k-ILH-differentiable, $1 \leq j \leq k$,
$$D^j\underline{\Phi} := \varprojlim D^j \Phi_\nu$$
is the jth derivative of $\underline{\Phi}$.

We now let M be a compact, connected, n-dimensional C^∞-manifold with boundary ∂M. We fix a C^∞-Riemannian metric $g = (g_{ij})_{i,j=1,...n}$ for which ∂M is totally geodesic and which satisfies
$$\int_M d\text{vol}(g) = 1.$$

g defines the Levi-Civita connection ∇. We let $\pi : TM \to M$ be the base point projection.

Let
$$\underline{V} := \{C^\infty\text{-sections } s : M \to TM \text{ with } s(p) \in T_p\partial M \text{ for } p \in \partial M,$$
$$\text{i.e. } s \text{ is tangential to } \partial M\}.$$

Then
$$\underline{V} = \varprojlim V_k,$$
where V_k is the completion of $\underline{V}$ with respect to the Sobolev norm $\|.\|_{H^k}$. We also let $\tilde{V}_k$ be the completion of $\underline{V}$ with respect to the Banach space norm $\|.\|_{C^k}$.

A crucial auxiliary result now is:

I.3: Harmonic maps and global structures

Lemma A.2.1: There exists an ε-neighborhood U^ε of 0 in $\tilde{V}_1$ such that for every $v \in U^\varepsilon$,
$$\phi_v : M \to M$$
$$p \to \exp_p v(p)$$
exists and is a diffeomorphism of M onto itself; in particular, $\phi_v(\partial M) = \partial M$.

Proof: Suppose ϕ_v exists; we want to estimate its derivative $d\phi_v$. This derivative can be represented by Jacobi fields in a standard way:
$$d\phi_v(X) = J_x(|v(p)|), \tag{A.2.1}$$
where $J_x(t)$ solves the Jacobi equation
$$J^{l''}(t) + R^l_{ijk} \dot{c}^j(t) \dot{c}^k(t) J^i(t) = 0, \tag{A.2.2}$$
where R^l_{ijk} is the curvature tensor of M and $\dot{c}$ is the tangent vector to the geodesic $\exp_p t \frac{v(p)}{|v(p)|}$, supposing this geodesic exists for $0 \le t \le \delta$, and J_X satisfies the initial conditions
$$J_X(0) = X, \; J'_X(0) = \nabla_X v. \tag{A.2.3}$$
If $X \ne 0$ and $\tilde{J}_X$ is the solution of (A.2.2) with initial conditions
$$\tilde{J}_X(0) = X, \; \tilde{J}'_X(0) = 0,$$
then
$$\tilde{J}_X(t) \ne 0$$
if $t \le i(M)$, where $i(M) > 0$ is the injectivity radius of M.

Since solutions of (A.2.2) depend continuously on the initial conditions, we conclude that, if $|X| = 1$ and $|\nabla v| \le \varepsilon'$, for some $\varepsilon' > 0$,
$$J_X(t) \ne 0 \quad \text{for} \quad 0 \le t \le \varepsilon'' \quad \text{for some} \quad \varepsilon'' > 0.$$

Since M is compact, ε' and ε'' can of course be chosen independently of $p \in M$. Thus, if $|v| \le \varepsilon''$, $|\nabla v| \le \varepsilon'$,
$$d\phi_v \ne 0.$$

By a continuity argument, we conclude that for all s with $0 \le s \le 1$, ϕ_{sv} is a diffeomorphism and in particular cannot map an interior point of M onto a boundary point. In particular, the geodesic arc $\exp_p t \frac{v(p)}{|v(p)|}$ exists for $0 \le t \le \varepsilon'$. The claim follows with $\varepsilon = \min(\varepsilon', \varepsilon'')$.

We now want to introduce a topology on D. Choose j large enough that $V_j \subset \tilde{V}_1$.

$D'_j := \{v \in V_j : \|v\|_{C^1} < \varepsilon\}$, where ε is given in Lemma A.2.2.
$D_j := \{\phi_v : v \in D'_j\}$
$\underline{\underline{\Phi}} : D'_j \to D_j$
$\underline{\underline{\Phi}}(v) := \phi_v$

As D'_j is an open subset of V_j, we can introduce a topology on D_j through $\underline{\underline{\Phi}}$.
We then put
$$\mathcal{D}^j := \cup\{D_j \phi, \quad \phi \in \mathcal{D}\},$$
thus introducing on $\mathcal{D}^j$ the topology of D_j. Finally the topology on D is given through
$$\mathcal{D} = \varprojlim \mathcal{D}^j.$$

We also put

$\mathcal{D}' := \{\phi \in \mathcal{D} : \text{ if in local coordinates}, \partial M \text{ is given as}$
$z^n = 0$, then
$\dfrac{\partial \phi^1}{\partial z^n} = \cdots \dfrac{\partial \phi^{n-1}}{\partial z^n} = 0\} \quad (n = \dim M)$.

$\mathcal{D}'$ is topologized in the same way as $\mathcal{D}$. Using the arguments of [O] and Lemma A.2.1, one can show

Theorem A.2.1:
(i) $\mathcal{D}$ and $\mathcal{D}'$ are strong ILH-Lie groups.
(ii) $\exp : \underline{V} \to D$ is a C^∞-ILH-mapping.

I.3 Harmonic maps and global structures

We first recall some notation. Let S be a compact oriented 2-dimensional differentiable manifold with boundary ∂S. Set $\mathcal{D}^l = \{\text{diffeomorphisms } \phi : S \to S \text{ of Sobolev class } H^l\}$. We let $\mathcal{D}_0^l$ be the subgroup of $\mathcal{D}^l$ of diffeomorphisms of S homotopic to the identity (we require that throughout the homotopy, ∂S is always mapped into itself). Since for compact surfaces homotopic diffeomorphisms are isotopic (Baer's theorem), and since such an isotopy can be performed in H^l (cf. [J2]), $\mathcal{D}_0^l$ is the connected component of the identity in $\mathcal{D}^l$. As a subgroup of $\mathcal{D}^l$, $\mathcal{D}_0^l$ also acts on the spaces $\mathcal{M}^k$ and $\mathcal{M}_{-1}^k$ defined in the preceding section.

We first consider the action

$$\mathcal{M}_{-1}^k \times \mathcal{D}_0^{l+1} \to \mathcal{M}_{-1}^l, \quad l \leq k. \tag{3.1}$$

We shall verify below that every $g \in \mathcal{M}_{-1}^l$ is orbit equivalent to some $g' \in \mathcal{M}_{-1}^\infty$, i.e. there exists $\phi \in \mathcal{D}_0^{l+1}$ with $\phi^* g \in \mathcal{M}_{-1}^\infty$. Consequently, the space of orbits is independent of l and k.

Definition 3.1: Let $p \geq 2$, where p is the genus of the Schottky double S^d of S. The Teichmüller space T_p is then the space of orbits of the action (3.1).

In order to define a differentiable structure on T_p compatible with the differentiable structures on the spaces $\mathcal{M}^k$, we shall use the concept of harmonic maps between surfaces.

Let Σ, Σ' be compact Riemann surfaces, possibly with boundary. Let $z = x + iy$ be a local conformal parameter on Σ, $u = u^1 + iu^2$ a local conformal parameter on Σ', and $\rho^2(u)dud\bar{u}$ a conformal metric on Σ'. Assume that $\partial \Sigma'$, if nonempty, is totally geodesic with respect to this metric. The <u>energy</u> of a map $u(z)$ of Sobolev class $H^1, u : \Sigma \to \Sigma'$, is defined as

$$E(u) = \int_\Sigma \rho^2(u)(u_z \bar{u}_{\bar{z}} + u_{\bar{z}} \bar{u}_z) dz d\bar{z}. \tag{3.2}$$

u is called <u>harmonic</u> if it is a critical point of the energy functional $E(u)$. Here, we look for critical points in the class of all maps that map $\partial \Sigma$ onto $\partial \Sigma'$. We shall need the following result that summarizes work of Eells, Sampson, Hartman, Schoen and Yau.

Theorem 3.1: Assume that the metric $\rho^2(u)dud\bar{u}$ has negative Gauss curvature. Then each $h : \Sigma \to \Sigma'$ is homotopic to a unique harmonic map $u : \Sigma \to \Sigma'$ satisfying the boundary condition (3.5) below. u minimizes the energy among all maps $(\Sigma, \partial\Sigma) \to (\Sigma', \partial\Sigma')$ homotopic to h. If h is a diffeomorphism, then so is u.

As a harmonic map, u satisfies the equation

$$u_{z\bar{z}} + \frac{2\rho_u}{\rho} u_z u_{\bar{z}} = 0, \qquad (3.3)$$

and this in the present case is equivalent to

$$\Phi dz^2 = \rho^2(u) u_z \bar{u}_z dz^2 \qquad (3.4)$$

being a holomorphic quadratic differential. Furthermore, if at $p \in \partial\Sigma$, $\frac{\partial}{\partial t}$ and $\frac{\partial}{\partial n}$ are tangential and normal vectors, respectively,

$$u_t \cdot u_n = 0. \qquad (3.5)$$

In terms of Φ this means that Φdz^2 is real on $\partial\Sigma$.

Finally, $\Phi dz^2 = 0$ if and only if u is holomorphic or antiholomorphic.

Proof: In [J1] and [J2], proofs of these results for closed surfaces are given. The present case is easily reduced to the case of closed surfaces by taking the Schottky doubles Σ^d, Σ'^d of Σ, Σ' respectively Σ'^d carries a hyperbolic metric which is invariant under a reflection leaving $\partial\Sigma'$ fixed so that $\partial\Sigma'$ is a closed geodesic w.r.t. this metric. By uniqueness, the corresponding harmonic map

$$u^d : \Sigma^d \to \Sigma'^d$$

is invariant when composed with the reflection of Σ^d and $\Sigma^{d'}$ simultaneously and thus restricts to a harmonic map

$$u : \Sigma \to \Sigma', \quad \text{mapping } \partial\Sigma \text{ onto } \partial\Sigma'.$$

(3.5) again follows from the fact that u^d is invariant under the above reflections. The regularity of u will be discussed in Lemmas 3.1 and 3.2 below. ∎

For the moment, we shall adopt the following point of view. We take Σ as fixed and Σ' as variable. By Poincaré's theorem, each Σ' carries a unique hyperbolic metric $\rho^2(u)dud\bar{u}$, and we shall identify the conformal structure with the corresponding hyperbolic metric. We take a differentiable reference surface S and write

$$\Sigma' = (S, \gamma) = (S, \rho^2(u)dud\bar{u})$$

I.3: Harmonic maps and global structures

where γ denotes the conformal (or hyperbolic) structure.

An important point is that this construction is equivariant with respect to the action of $\mathcal{D}_0$; if $i : \Sigma' \to \Sigma''$ is an isometry, then $u : \Sigma \to \Sigma'$ is harmonic if and only if $i \circ u : \Sigma \to \Sigma''$ is harmonic. This follows from the invariance of (3.3) under isometries.

If now $\Phi \in \mathcal{D}_0^l$, then

$$\phi : \left(S, \phi^*(\rho^2(u)dud\bar{u})\right) \to \left(S, \rho^2(u)dud\bar{u}\right)$$

is an isometry. Consequently, if $u : \Sigma \to \left(S, \rho^2(u)dud\bar{u}\right)$ is harmonic, then

$$\phi^{-1} \circ u : \Sigma \to \left(S, \phi^*(\rho^2(u)dud\bar{u})\right)$$

is likewise harmonic. Furthermore Φdz^2, as given in (3.4), remains invariant. We define

$Q(\Sigma) := \{\text{holomorphic quadratic differentials on } \Sigma \text{ which are real on } \partial\Sigma\}$.

The preceding consideration implies that we obtain a map

$$q(\Sigma) : T_p \to Q(\Sigma)$$

by taking the holomorphic quadratic differential associated with the harmonic map $u : \Sigma \to \Sigma'$ homotopic to the identity of the underlying reference surface S. The following result is due to S. Wolf [Wo]; cf. also [J2].

Theorem 3.2: For each fixed Σ, the map $q(\Sigma)$ is bijective. ∎

The underlying construction is the following. Given $\Phi dz^2 \in Q(\Sigma)$, one has to construct a hyperbolic metric $\rho^2(u)dud\bar{u}$ and a harmonic map $u = u(z)$ so that the pullback of the metric under u,

$$\rho^2(u)u_z\bar{u}_z dz^2 + \rho^2(u)(u_z\bar{u}_{\bar{z}} + \bar{u}_z u_{\bar{z}})dzd\bar{z} + \rho^2(u)\bar{u}_{\bar{z}}u_{\bar{z}}d\bar{z}^2, \qquad (3.6)$$

satisfies

$$\rho^2(u)u_z\bar{u}_z = \Phi. \qquad (3.7)$$

For this one puts, using the hyperbolic metric $\lambda^2(z)dzd\bar{z}$ on Σ,

$$H := \frac{\rho^2(u)}{\lambda^2(z)}u_z\bar{u}_{\bar{z}}, \quad L := \frac{\rho^2(u)}{\lambda^2(z)}\bar{u}_z u_{\bar{z}} \qquad (3.8)$$

(in invariant notation

$$H = \|\partial u\|^2, \; L = \|\bar{\partial} u\|^2).$$

One observes
$$H \cdot L = \frac{1}{\lambda^4}\Phi\bar{\Phi} \tag{3.9}$$
and computes, for u harmonic,
$$\Delta \log H = -2 + 2(H - L), \tag{3.10}$$
where Δ is the Laplace–Beltrami operator for $\lambda^2(z)dzd\bar{z}$, and, as stated before, both metrics are hyperbolic. Given Φ, one solves (3.9) and (3.10) for H and L defined by (3.8); the required hyperbolic metric is then
$$\Phi dz^2 + \lambda^2(z)(H+L)dzd\bar{z} + \bar{\Phi}d\bar{z}^2, \tag{3.11}$$
and we can then uniformize this metric as
$$\rho^2(u)dud\bar{u},$$
which determines u. Moreover (cf. [J2]) as a consequence of elliptic regularity theory:

Theorem 3.3: The transition functions
$$q(\Sigma_2) \circ q(\Sigma_1)^{-1} : Q(\Sigma_1) \to Q(\Sigma_2)$$
are of class C^∞. ■

Theorem 3.3 defines a canonical (because independent of the choice of Σ) differentiable structure on T_p. We shall now compare this structure to the differentiable structure of $\mathcal{M}^l$.

We want to address the regularity of u in terms of the image metric. For this, it is convenient to rewrite (3.3) in real form. We also want to consider the case of a general image metric (g_{ij}), not necessarily given in isothermal parameters. Writing $u = (u^1, u^2)$, $z = z^1 + iz^2$, the energy of u becomes
$$E(u) = \frac{1}{2}\int_\Sigma g_{ij}(u(z))u^i_{z^\alpha}u^j_{z^\alpha}dzd\bar{z}, \tag{3.12}$$
and the Euler–Lagrange equations which u as a critical point of E has to satisfy are then
$$\Delta u^i + \Gamma^i_{jk}(u)u^j_{z^\alpha}u^k_{z^\alpha} = 0, \, i = 1, 2, \tag{3.13}$$
where $\Gamma^i_{jk} = \frac{1}{2}g^{il}(g_{jl,k} + g_{kl,j} - g_{jk,l})$ are the usual Christoffel symbols. We quote the following result (cf. [J2] for proofs and references):

Lemma 3.1: If u is a minimum of $E(u)$ (for a continuous (g_{ij})), then u is continuous. (By uniqueness this is the case for example if the image metric

I.3: Harmonic maps and global structures

g_{ij} has nonpositive curvature). If $g_{ij} \in C^{k,\alpha}, 1 \leq k \leq \infty, 0 < \alpha < 1$ and u is a continuous weak solution of (3.13), then $u \in C^{k+1,\alpha}$.

∎

We shall now prove:

Lemma 3.2: Let u be a continuous weak solution of (3.13), for example a minimum of $E(u)$. If

$$(g_{ij}) \in H^l, \text{ for } l \geq 3,$$

then

$$u \in H^{l+1}.$$

Proof: We shall need the general Sobolev spaces

$$H^{k,p} := \Big\{ f : \Sigma \to \mathbb{R}, (\sum_{|\alpha|=1}^{k} \int_{\Sigma} |D^\alpha f|^p)^{\frac{1}{p}} < \infty,$$

where $D^\alpha f$ is a weak derivative of order $\alpha \Big\}$.

$u \in H^{k,p}$ then will mean $u^1, u^2 \in H^{k,p}$ in local coordinates. In our previous notation,

$$H^l = H^{l,2}.$$

The Sobolev embedding theorem says that $H^{1,p}$ embeds for $p < 2$ into

$$L^{\frac{2p}{2-p}},$$

and for $p > 2$ into

$$C^{0,1-\frac{2}{p}}.$$

We start with the case $l = 3$. Then $g_{ij} \in H^{3,2} \Rightarrow \Gamma^i_{jk} \in H^{2,2} \Rightarrow \Gamma^i_{jk} \in C^\beta$, for some $\beta \in (0,1)$, by the Sobolev embedding theorem. Lemma 3.1 yields $u \in C^{2,\alpha}$ for some $\alpha \in (0,1)$. In particular, $u \in H^{2,2}$ and $\nabla u \in H^{1,2}$. We now observe that

$$f \in H^{1,2} \Rightarrow f^2 \in H^{1,2-\varepsilon} \text{ for all } \varepsilon \in (0,1),$$

again with the help of the Sobolev embedding theorem. Since $\Gamma^i_{jk} \in H^{1,p}$ for every $p < \infty$, again by Sobolev, the same is true for $\Gamma^i_{jk} \circ u$, and then

$$\Gamma^i_{jk}(u) u^j_{z^\alpha} u^k_{z^\alpha} \in H^{1,2-\delta} \text{ for all } \delta \in (0,1).$$

We conclude from (3.13) with the help of a classical Calderón–Zygmund theorem

$$u \in H^{3,2-\delta} \text{ for all } \delta \in (0,1),$$

hence $\nabla u \in H^{2,2-\delta}$. Also if f and g are in $H^{2,2-\delta}$ for all $\delta \in (0,1)$ then the same holds for fg, again via the Sobolev embedding theorem. Similarly,

$$f, g \in H^{2,2} \Rightarrow fg \in H^{2,2} \tag{3.14}$$

Thus
$$\Gamma^i_{jk}(u) u^j_{z^\alpha} u^k_{z^\alpha} \in H^{2,2-\eta} \text{ for all } \eta \in (0,1),$$
and again by Calderón–Zygmund, as u solves (3.13),

$$u \in H^{4,4-\eta} \text{ for all } \eta \in (0,1).$$

But then $|\nabla u|^2 \in H^{2,2}$ by Sobolev, and, using (3.14),

$$\Gamma^i_{jk}(u) u^j_{z^\alpha} u^k_{z^\alpha} \in H^{2,2}$$

and finally again by Calderón–Zygmund

$$u \in H^{4,2}.$$

This proves the claim for $l = 3$. For $l \geq 4$, we may assume by induction $u \in H^{l,2}$. Then in the same way as before

$$\Gamma^i_{jk}(u) u^j_{z^\alpha} u^k_{z^\alpha} \in H^{l-1,2},$$

and by Calderón–Zygmund

$$u \in H^{l+1,2}.$$ ∎

Let $\mathcal{M}_{neg}$ be the subspace of $\mathcal{M}$ consisting of negatively curved metrics.

Lemma 3.3: The map

$$\mathcal{M}_{neg} \to \mathcal{D}_0$$

mapping g onto the harmonic map $u : (S, g_0) \to (S, g)$, g_0 fixed, homotopic to the identity of S, is of class C^∞.

Remark: The corresponding map $\mathcal{M}^l_{neg} \to \mathcal{D}^{l+1}_0$ is only continuous, not differentiable.

Proof: Let g_t be a smooth family of metrics and $u_t : (S, g_0) \to (S, g_t)$ be the corresponding harmonic maps. Differentiating

$$\Delta u^i_t + {}^t\Gamma^i_{jk}(u_t)(u^j_t)_{z^\alpha}(u^k_t)_{z^\alpha} = 0$$

with respect to t at $t = 0$ and denoting derivatives at $t = 0$ by a dot and omitting a subscript 0, we get

$$\Delta \dot{u}^i + 2\Gamma^i_{jk} \dot{u}^j u^k_{z^\alpha} + \dot{\Gamma}^i_{jk} u^j_{z^\alpha} u^k_{z^\alpha} + \Gamma^i_{jk,l} \dot{u}^l u^j_{z^\alpha} u^k_{z^\alpha} = 0, i = 1, 2. \tag{3.15}$$

I.3: Harmonic maps and global structures

Given $u = u_0$ and $g = g_0$ and the variation $\dot{g}$, this is a linear elliptic equation for $\dot{u}$. The uniqueness result for harmonic maps into negatively curved metrics implies that the solution $\dot{u}$ is likewise unique, and elliptic regularity theory then implies the claim.

∎

The fact that the corresponding map $\mathcal{M}_{neg}^l \to \mathcal{D}_0^{l+1}$ is not differentiable follows from the presence in (3.15) of the derivative $\Gamma^i_{jk,l}$ which is only of class H^{l-2} for $g \in \mathcal{M}^l$, so that the solution $\dot{u}$ is only in H^l, not in H^{l+1}. We can now prove the following strengthening of a result of Earle and Eells [EE] (cf. also [T1,2] for a similar result):

Theorem 3.4: $\mathcal{M}_{-1}$ is C^∞-diffeomorphic to $T_p \times \mathcal{D}_0$.

Proof: $g \in \mathcal{M}_{-1}$ defines a hyperbolic metric on our reference surface S. As before, we let Σ be a fixed Riemann surface, i.e. a fixed conformal structure on S, and take the harmonic map

$$u_g : \Sigma \to (S, g)$$

homotopic to the identity of S. u defines a holomorphic quadratic differential $\Phi_g dz^2$ on Σ, and we define

$$F : \mathcal{M}_{-1} \to T_p \times \mathcal{D}_o$$
$$g \to (\Phi_g, (u_g)^{-1}). \tag{3.16}$$

Since by Lemma 3.3 $g \to u_g$ as a map from $\mathcal{M}_{-1}$ to $\mathcal{D}_0$ is of class C^∞, so is F. (We do not know yet that $\mathcal{M}_{-1}$ is a manifold, but F can be defined in a neighborhood of $\mathcal{M}_{-1}$ in $\mathcal{M}$, for example on $\mathcal{M}_{neg}$, and is C^∞ there, as a consequence of uniqueness of harmonic maps in this case). The inverse map F^{-1} is given as follows $\Phi dz^2 \in Q(\Sigma)$ defines a C^∞-metric

$$g(\Phi) := \Phi dz^2 + \lambda^2 (H + L) dz d\bar{z} + \bar{\Phi} d\bar{z}^2 \tag{3.17}$$

as explained after Theorem 3.2. Then, using the identification $q(\Sigma) : T_p \to Q(\Sigma)$ of Theorem 3.2,

$$F^{-1} : T_p \times \mathcal{D}_0 \to \mathcal{M}_{-1}$$
$$(\Phi dz^2, \phi) \to ((\phi^{-1})^* g(\Phi)). \tag{3.18}$$

F^{-1} is of class C^∞ because $g(\Phi) \in C^\infty$. It is clear from the constructions that both F and F^{-1} are bijective and indeed inverse to each other. In the same way, one establishes homeomorphisms

$$F^l : \mathcal{M}_{-1}^l \to T_p \times \mathcal{D}_0^{l+1}, l \geq 3,$$

using Lemma 3.2. ∎

Another consequence of Lemma 3.2 is:

Corollary 3.1: Every $g \in \mathcal{M}_{-1}^l$ is equivalent to some $g' \in \mathcal{M}_{-1}$ under the action of $\mathcal{D}_0^{l+1}$.

Proof: As before, we construct the harmonic map $u : \Sigma \to (S, g)$, and by Lemma 3.2,
$$u \in \mathcal{D}_0^{l+1}.$$

As in the proof of Theorem 3.4, u defines $\Phi dz^2 \in Q(\Sigma)$ which in turn yields the smooth metric $g(\Phi)$ of (3.17). $g(\Phi)$ is conformally equivalent to g, and the claim follows. ∎

We now turn to global decompositions of $\mathcal{M}^l$. $g \in \mathcal{M}^l, l \geq 3$, can be uniformized by a smooth hyperbolic metric g_{-1} on S, by Poincaré's uniformization theorem and its extension to the case of continuous metrics; see [J2] for details. The corresponding conformal map

$$u : (S, g_{-1}) \to (S, g)$$

is then harmonic (a conformal map obviously satisfies (3.3)), hence of class H^{l+1} by Lemma 3.2. $u^*(g)$ is then also of class H^l, and therefore differs from g_{-1} by a positive function of class H^l. Conversely, multiplying any $g' \in \mathcal{M}_{-1}^l$ by a positive H^l function yields an element $g \in \mathcal{M}^l$, for $l \geq 3$. We define

$$C^l := \{f \in H^l(S, \mathbb{R}), f > 0\}, \quad C := C^\infty$$

and conclude from Theorem 3.4 and the preceding construction the validity of:

Theorem 3.5: $\mathcal{M}^l$ is homeomorphic to $T_p \times \mathcal{D}_0^{l+1} \times C^l$, and $\mathcal{M}$ is diffeomorphic to $T_p \times \mathcal{D}_0 \times C$. ∎

Moreover $\mathcal{M}^l$ is convex, hence starshaped with respect to any of its points, hence contractible, and we then have:

Corollary 3.2: $\mathcal{M}_{-1}^l, \mathcal{D}_0^{l+1}$, and T_p are contractible for $l \geq 3$. ∎

Of course, we knew the contractibility of T_p already from Theorem 3.2.

We now discuss gauge fixing. A <u>gauge</u> is a section

$$\sigma : T_p \longrightarrow \mathcal{M}$$
$$t \longrightarrow g_t$$

I.3: Harmonic maps and global structures

of the bundle $\pi : \mathcal{M} \to T_p$ constructed in Theorem 3.5. Above, we have already constructed a global gauge, namely the so-called harmonic gauge (first established in [EE])

$$\sigma_h : T_p \longrightarrow \mathcal{M}_{-1} \qquad (3.19)$$
$$t \longrightarrow g_t$$

given as follows. For $g \in \mathcal{M}_{-1}$, let $\Phi dz^2 = \pi(g) \in Q(\Sigma)$ as in Theorem 3.4, and let

$$u_g : \Sigma \to (S, g)$$

be the harmonic map. Then

$$\sigma_h(\Phi) := (u_g)^*(g) \in \mathcal{M}_{-1}. \qquad (3.20)$$

This is independent of the choice of $g \in \pi^{-1}(\Phi dz^2)$. Namely, for $g \in \mathcal{M}_{-1}$, any other $g' \in \pi^{-1}(\pi(g))$ is of the form $\phi^* g, \phi \in \mathcal{D}_0$. As remarked before Theorem 3.2, we have

$$u_g = \phi \circ u_{\phi^* g}, \qquad (3.21)$$

hence

$$(u_{\phi^* g})^*(\phi^* g) = (u_g)^*(\phi^{-1})^*(\phi^* g) = (u_g)^* g, \qquad (3.22)$$

proving that $\sigma_h(\Phi)$ is independent of the choice of g.

Local gauges can also be constructed as follows (cf. [T1,2]). Let $g \in \mathcal{M}_{-1}$, and let $h'' \in T_g \mathcal{M}$ be tracefree and divergencefree and satisfy (3.26). If $|h''| < \varepsilon$ for sufficiently small ε, then $g + h''$ is positive definite, hence $g + h'' \in \mathcal{M}$. Again by Poincaré's uniformization theorem, there exists a positive C^∞-function $\lambda^2(z)$ with

$$\lambda^2(z)(g + h'') \in \mathcal{M}_{-1}.$$

As h'' corresponds to a holomorphic quadratic differential on (S, g) (see I.2),

$$h'' \to \lambda^2(z)(g + h'')$$

is then a local gauge.

There are certain odds and ends left over. Traditionally, Teichmüller space is defined as the space of complex structures on S divided by the action of $\mathcal{D}_0$. Of course, by Poincaré's theorem, complex structures and hyperbolic metrics are in bijective correspondence to each other, and it is not difficult to see from the formulae below that the spaces are actually diffeomorphic. This diffeomorphism, however, effects a change of type of tensors.

Namely, a metric is given by a symmetric positive definite C^∞ (or H^l) (0,2) tensor. On the other hand, on a Riemann surface a complex structure is the same as an almost complex structure which in turn is given by a C^∞ (or H^l) $(1,1)$ tensor J, namely for each $z \in S$

$$J(z) : T_z S \to T_z S \text{ with } J^2(z) = -id_{|T_z S}.$$

A metric (g_{ij}) defines an almost complex structure J via

$$J_i^k = g^{kj}(\det g)^{\frac{1}{2}}(\delta_{1i}\delta_{2j} - \delta_{1j}\delta_{2i}),$$

and conversely (J_i^k) determines (g_{ij}) up to a conformal factor. From this point of view, namely looking at Teichmüller spaces as equivalence classes of complex structures, the tangent space of T_p should consist of $(1,1)$ tensors, whereas $(0,2)$ tensors should yield cotangent vectors. Moreover $Q(\Sigma)$, the space of holomorphic quadratic differentials on Σ which are real on $\partial\Sigma$, is the cotangent space of T_p at Σ. The tangent space of T_p at Σ is the space $H(\Sigma)$ of harmonic Beltrami differentials on Σ: if $\Phi dz^2 \in Q(\Sigma)$, and $\lambda^2(z)dzd\bar{z}$ is a conformal metric, then

$$\frac{1}{\lambda^2(z)}\bar{\Phi}(z)d\bar{z} \otimes \frac{\partial}{\partial z} \in H(\Sigma),$$

and conversely. In real notation, if h''_{ik} is tracefree and divergencefree,

$$h_i^{''j} = g^{jk}h''_{ik}$$

is the corresponding tangent vector.

The preceding considerations are only valid if the genus p of the Schottky double of our surface is at least 2. Let us briefly discuss the remaining cases of surfaces with boundary, namely the unit disk D and an annulus A. It follows from the Riemann mapping theorem that D has only one complex structure so that the corresponding Teichmüller space is trivial. It is also a classical result that the complex structures on A are parametrized by the real numbers. Of course, both these statements hold up to the action of $\mathcal{D}_0$. Let us briefly show that in both these cases, $\mathcal{D}_0$, instead of being contractible, has S^1 as a strong deformation retract.

In the case of D, one first retracts $\mathcal{D}_0$ onto the space of those diffeomorphisms that fix 0. Then, for each $r, 0 < r \leq 1$, and each such diffeomorphism ϕ, one has a unique conformal diffeomorphism $k(\phi, r)$ from $\phi(B(0, r))$ onto D with $k(\phi, r)(0) = 0$ and

$$\frac{grad\, k(\phi, r)(0)}{|grad\, k(\phi, r)(0)|} = \frac{grad\, \phi(0)}{|grad\, \phi(0)|}.$$

In this way, one can retract $\mathcal{D}_0$ onto the conformal automorphisms of D fixing 0. This latter space is S^1. Of course, the important point in the previous argument is that the normalized conformal map of the Riemann mapping theorem depends continuously on the domain. The case of an annulus is handled by a similar argument.

I.4 Cauchy–Riemann operators

Let S be a differentiable surface and let Σ denote a conformal structure on S so that Σ can be seen as a Riemann surface. Let g be a conformal metric on Σ. For what follows it will be convenient to rewrite things in complex notation. The metric is then given by:

$$\rho^2(z)dzd\bar{z} \tag{4.1}$$

with a conformal parameter $z = z^1 + iz^2 = x + iy$.

Let K be the canonical bundle of our Riemann surface Σ; sections of K are of the form $f(z)dz$, where $f : \Sigma \to \mathbb{C}$. The inverse K^{-1} of K has sections $f(z)\frac{\partial}{\partial z}$, while $\bar{K}$ has sections $f(z)d\bar{z}$. Now $K \otimes \bar{K}$ is a trivial bundle, a global nonvanishing section being given by the metric

$$\rho^2(z)dzd\bar{z}.$$

Thus, we have an isomorphism

$$K^{-1} \to \bar{K}, \quad \frac{\partial}{\partial z} \to \rho^2 d\bar{z}. \tag{4.2}$$

If $V^1 \frac{\partial}{\partial x} + V^2 \frac{\partial}{\partial y}$ is a vector field, we form the associated $(1,0)$ vector field

$$V\frac{\partial}{\partial z} = (V^1 + iV^2)\frac{\partial}{\partial z}.$$

The covariant derivative of $V\frac{\partial}{\partial z}$ in the $\frac{\partial}{\partial \bar{z}}$-direction is given by

$$\bar{\partial}(V\frac{\partial}{\partial z}) = V_{\bar{z}}d\bar{z} \otimes \frac{\partial}{\partial z} = \rho^2 V_{\bar{z}}d\bar{z}^2 \tag{4.3}$$

under the isomorphism (4.2). As $\bar{\partial}$ in (4.3) maps K^{-1} into $K^{-1} \otimes \bar{K} \cong K^{-2}$, we identify $\bar{\partial}$ with $\bar{\partial}_{-1}$ and write

$$\bar{\partial}_{-1}(V\frac{\partial}{\partial z}) = \rho^2 V_{\bar{z}}d\bar{z}^2 = \frac{1}{\rho^2}V_{\bar{z}}(\frac{\partial}{\partial z})^2. \tag{4.4}$$

Likewise, we have operators

$$\bar{\partial}_n : K^n \to K^{n-1} \tag{4.5}$$

and the adjoint of $\bar{\partial}_n$ (with suitable boundary conditions) is given by

$$\bar{\partial}_n^* W = -(\rho^2)^{n-1}\frac{\partial}{\partial z}((\frac{1}{\rho^2})^{n-1}W)dz$$

$$\bar{\partial}_n^* : K^{n-1} \to K^n.$$

I.4: Cauchy-Riemann operators

In particular,

$$\bar{\partial}_z^*(\bar{V}\frac{\partial}{\partial \bar{z}}) = \bar{\partial}_z^*(\rho^2 \bar{V} dz) = -\rho^2 \bar{V}_z dz^2. \tag{4.6}$$

In the following we introduce operators, built up from the Cauchy-Riemann operator $\bar{\partial}_n$, which will play an important role in the functional quantization of bosonic strings. Let

$$\triangle_g := \frac{1}{(\det g)^{\frac{1}{2}}} \frac{\partial}{\partial z^i} \left((\det g)^{\frac{1}{2}} g^{ij} \frac{\partial}{\partial z^j} \right) \tag{4.7}$$

be the Laplace–Beltrami operator corresponding to the metric g. It can be expressed in terms of $\bar{\partial}_0, \bar{\partial}_0^*$ as:

$$\triangle_g = -2\partial_0^* \bar{\partial}_0. \tag{4.8}$$

Let now P_g be the operator which maps vector fields into symmetric tracefree 2×2 tensors defined by

$$P_g(V)_{ij} = 2G_{ij}^{kl} g_{km} (\nabla_{\frac{\partial}{\partial z^l}} V)^m \tag{4.9}$$

for a vector field V. We recall that $G_{ij}^{kl} \equiv \frac{1}{2} \left(\delta_i^k \delta_j^l + \delta_i^l \delta_j^k - g_{ij} g^{kl} \right)$. We also remark that $P_g(V)_{ij} = (\nabla_g V)_{ij} - \frac{1}{2} tr_g(\nabla_g V) g_{ij}$ where $(\nabla_g V)_{ij} = \nabla_i V_j + \nabla_j V_i$ is the Lie derivative of V with respect to g, and $tr_g h = g^{ij} h_{ij}$ is the trace of a tensor h with respect to the metric g. Using isothermal coordinates and (4.1), we can write:

$$P_g(V) = \rho^2 \bar{V}_z dz^2 + \rho^2 V_{\bar{z}} d\bar{z}^2.$$

If $z = \phi(w)$ is a diffeomorphism, then the metric $\rho^2(z) dz d\bar{z}$ is pulled back to

$$\rho^2(\phi(w)) d\phi(w) d\bar{\phi}(w) = \rho^2(\phi(w))(\phi_w \bar{\phi}_w dw^2 + (\phi_w \bar{\phi}_{\bar{w}} + \phi_{\bar{w}} \bar{\phi}_w) dw d\bar{w} + \phi_{\bar{w}} \bar{\phi}_{\bar{w}} d\bar{w}^2).$$

Thus, if $z = \phi_t(w)$ is a family of diffeomorphisms with $\phi_0 = \text{id}$, $\dot{\phi}_0 = V$, then

$$\frac{d}{dt}(\rho^2(\phi_t(u)) d\phi d\bar{\phi})_{|t=0} =$$
$$((\rho^2(z) V(z))_z + (\rho^2(z) \bar{V}(z))_{\bar{z}}) dz d\bar{z} + \rho^2(z) \bar{V}(z)_z dz^2 + \rho^2(z) V(z)_{\bar{z}} d\bar{z}^2.$$

The generator P_g is then the tracefree part of this expression and can be expressed in terms of $\bar{\partial}, \bar{\partial}^*$ as:

$$P_g(V) = \bar{\partial}_{-1} V \frac{\partial}{\partial z} \oplus \bar{\partial}_2^* \bar{V} \frac{\partial}{\partial \bar{z}} \in \bar{K}^2 \oplus K^2 \tag{4.10}$$

I.4: Cauchy-Riemann operators

(see (4.4), (4.6)). Its formal adjoint with respect to inner products on vector fields and 2×2 tensors induced by the metric g_{ij} reads:

$$P_g^* h = \frac{1}{(\det g)^{\frac{1}{2}}} \frac{\partial}{\partial z^j} \left((\det g)^{\frac{1}{2}} g^{jk} h_{ik} \right) \quad (4.11)$$

when applied to a symmetric tracefree 2×2 vector h. It can be expressed in terms of ∂, ∂^* as:

$$P_g^* h = \partial_2(h_1 dz^2) \oplus \partial_{-1}^*(h_2 d\bar{z}^2) \quad (4.12)$$

with $h = h_1 dz^2 \oplus h_2 d\bar{z}^2$.

In order to discuss self-adjointness of the operators $\triangle_g$ and $P_g^* P_g, P_g P_g^*$, we need to specify the boundary conditions. We first discuss the boundary conditions for the operators $\bar{\partial}_0^* \bar{\partial}_0$ and $\bar{\partial}_0 \bar{\partial}_0^*$. For a complex-valued function $f = f^1 + if^2$, we require

$$f^1 = 0 \text{ on } \partial \Sigma. \quad (4.13)$$

For reasons of holomorphicity, our second boundary condition is a Neumann condition:

$$f_y^2 = 0 \text{ on } \partial \Sigma, \text{ locally given as } y = 0. \quad (4.14)$$

With these boundary conditions,

$$\bar{\partial}_0^* \bar{\partial}_0 : K^0 \to K^0$$

becomes self-adjoint; the kernel of $\bar{\partial}_0$ then consists of the imaginary constants:

$$\ker \bar{\partial}_0 = i\mathbb{R}. \quad (4.15)$$

In the same manner as before, we write an element of K^{-1} as

$$g(z)d\bar{z} = (g^1(z) + ig^2(z))d\bar{z}$$

and

$$\bar{\partial}_0^*(g(z)d\bar{z}) = -\frac{1}{\rho^2} g_z dz d\bar{z}.$$

The corresponding boundary conditions for $g(z)$ are

$$g^1 = 0, g_y^2 = 0 \text{ on } \partial \Sigma \text{ (locally given as } y = 0), \quad (4.16)$$

and thus $\bar{\partial}_0 \bar{\partial}_0^*$ becomes selfadjoint. The kernel of $\bar{\partial}_0^*$ consists of complex conjugates of holomorphic 1-forms which are imaginary on $\partial \Sigma$; again, if g is antiholomorphic, $g^1 = 0$ implies $g_y^2 = 0$ on $\partial \Sigma$ by Cauchy-Riemann equations.

Let us now discuss the boundary conditions for $P_g^* P_g$ and $P_g P_g^*$. For simplicity of notation, we shall use again isothermal coordinates, i.e.

$$g_{ij}(z) = \rho^2(z)\delta_{ij} \quad (4.17)$$

with $z = x + iy$, where x is tangential and y is normal to the boundary, ∂S; thus, we can assume that ∂S is locally given as $y = 0$. We write again for vector fields

$$V = V^1 \frac{\partial}{\partial x} + V^2 \frac{\partial}{\partial y} \qquad (4.18)$$

and for tracefree symmetric 2×2 tensor

$$h = \begin{pmatrix} h_{11} & h_{12} \\ h_{12} & h_{22} \end{pmatrix} \qquad (4.19)$$

with $h_{22} = -h_{11}$. Since all diffeomorphisms have to map ∂S into itself, we require one boundary condition, namely

$$V^2 = 0 \quad \text{for} \quad y = 0 \qquad (4.20)$$

(cf. (2.21)). On the other hand, in order to have the equality

$$\int h_{ij} P_g(V)_{ij} \frac{1}{\rho^2} dz = -\int \frac{\partial}{\partial z^j}(h_{ij}) V^i dz (z^1 = x, z^2 = y) \qquad (4.21)$$

we require

$$h_{12} = 0 \quad \text{for} \quad y = 0 \qquad (4.22)$$

(cf. (2.22)).

(4.20) and (4.22), then, are our boundary conditions. When expressed in terms of V, (4.22) becomes, with

$$h_{ij} = P_g(V)_{ij} = g_{ij,k} V^k + \rho^2 V^i_{z^j}, \qquad (4.23)$$

the condition

$$\frac{\partial}{\partial y} V^1 = 0 \quad \text{for} \quad y = 0. \qquad (4.24)$$

Conversely, with

$$V^i = \frac{\partial}{\partial z^j} h_{ij} =: h_{ij,j} \qquad (4.25)$$

(4.20) becomes

$$h_{22,2} = 0 \quad \text{for} \quad y = 0 \qquad (4.26)$$

which, since h is tracefree, is equivalent to

$$h_{11,2} = 0 \quad \text{for} \quad y = 0. \qquad (4.27)$$

Thus, we require the pair of boundary conditions

$$V^2 = 0, V^1_y = 0 \quad \text{for} \quad y = 0 \qquad (4.28)$$

I.4: Cauchy-Riemann operators

or equivalently,
$$h_{12} = 0, h_{11,2} = 0 \quad \text{for} \quad y = 0, \tag{4.29}$$
i.e. one Dirichlet and one Neumann condition for the tracefree components. One possible interpretation of these conditions is the following (motivations for the mixed boundary conditions are also discussed in [Jas2]).
We restrict the class of admissible metrics to those for which ∂S is geodesic and we parameterize ∂S, locally given as $y = 0$, as $x = x(t)$. Dots will denote t-derivatives. The condition that $x(t)$ is geodesic is then
$$\ddot{x}(t) + \Gamma^1_{11}\dot{x}(t)\dot{x}(t) = 0 \tag{4.30}$$
and
$$\Gamma^2_{11}\dot{x}(t)\dot{x}(t) = 0, \tag{4.31}$$
since $y(t) \equiv 0$.

We continue to use isothermal coordinates. While (4.30) is just a condition on the parametrization $x(t)$, (4.31) means
$$-\frac{1}{2}g^{22}g_{11,2} = 0,$$
i.e.
$$g_{11,2} = 0 \quad \text{for} \quad y = 0. \tag{4.32}$$
If we look at a tracefree variation $g_{ij}(\tau)$, with $\frac{d}{d\tau}g_{ij}|_{\tau=0} = h_{ij}$, then the requirement that (4.31) is preserved means, using $g_{12} = 0$,
$$0 = \frac{1}{2}(h^{22}(2g_{12,1} - g_{11,2}) + h^{21}g_{11,1} + g^{22}(2h_{12,1} - h_{11,2}))$$
$$= \frac{1}{2}(h^{21}g_{11,1} + g^{22}(2h_{12,1} - h_{11,2})), \tag{4.33}$$
since $g_{12} = 0$ and by (4.32). This condition, however, is implied by (4.29).

Likewise, (4.28) can be interpreted as a symmetry condition; if ∂S is totally geodesic, then S can be doubled across ∂S to obtain a closed surface S^d with a symmetric metric, and (4.28) means that V extends as a $C^{1,1}$ vector field to S^d. Vector fields V satisfying (4.28) generate diffeomorphisms that preserve the normal direction along ∂S, in addition to mapping ∂S into itself. The considerations of section I.3 are valid for this restricted class of metrics (i.e. those for which ∂S is totally geodesic) and diffeomorphisms (i.e. those preserving the normal direction along ∂S) without essential changes. The relevant remark is that the harmonic diffeomorphism between hyperbolic surfaces in section I.3 extends to a harmonic diffeomorphism between the Schottky doubles, and is equivariant with respect to the corresponding reflections in domain and image, and of course the hyperbolic metrics as restrictions of the metrics on the Schottky doubles always fulfil the requirement that the boundary becomes geodesic.

I.5 Zeta-function and heat-kernel determinants of an operator

Let A be a Hermitian matrix whose eigenvalues $\lambda_1, ..., \lambda_N$ are positive numbers. Introducing the function

$$\zeta_{A,N}(s) := \sum_1^N \frac{1}{\lambda_n^s}, \; s \in \mathbb{C}, \tag{5.0}$$

the determinant of A is recovered by the formula

$$\det A = e^{-\zeta'_{A,N}(0)} \tag{5.1}$$

where $\zeta'_{A,N}(0) := \frac{d}{ds}\zeta_{A,N}(s)\,|_{s=0}$. The infinite-dimensional generalization of this scheme consists in considering A as a nonnegative self-adjoint operator acting on a Hilbert space $\mathcal{H}$, with spectrum given by eigenvalues $\{\lambda_n\}_{n=1}^\infty$. By definition

$$\zeta_A(s) := {\sum_n}' \frac{1}{\lambda_n^s}, \; s \in \mathbb{C} \tag{5.2}$$

is the zeta-function of A ($\sum'$ means that we are summing only over the nonzero eigenvalues). We shall restrict ourselves to the case, also relevant for the applications, where A is an elliptic differential operator of degree d on a compact Riemannian manifold M of dimension m. We remark that in this case the eigenvalues of A accumulate to infinity so the formal determinant of the operator, given by the product of all its eigenvalues, diverges. Nevertheless the sum (5.2) is absolutely convergent for $Re(s) > \frac{m}{d}$ and therefore $\zeta_A(s)$ is holomorphic in this region of the plane. This follows from the classical asymptotic formula

$$\lambda_n \sim n^{\frac{d}{m}} \quad \text{as } n \to \infty \tag{5.3}$$

discovered by H. Weyl in the case of the Laplacian ($d = 2$) (see e.g. [Cha]; see [Shu] for generalizations). In other words when A is positive then A^{-s} has a finite trace for $Re(s)$ sufficiently large. The same result holds also in the more general case of a positive elliptic pseudodifferential operator on M ([See] and [Ta]).

It is well known that $\zeta_A(s)$ admits a meromorphic continuation to the whole plane which is regular at $s = 0$. We shall sketch the argument following [G1,2]. Let $\Gamma(s)$ denote the usual Γ function so that

$$\Gamma(s) = \int_0^\infty dt \, t^{s-1} e^{-t} \tag{5.4}$$

for $Re(s) > 0$. Assuming $Re(\lambda) > 0$ we have the equality

$$\Gamma(s)\lambda^{-s} = \int_0^\infty dt \, t^{s-1} e^{-t\lambda}. \tag{5.5}$$

I.5: Zeta-function and heat-kernel determinants of an operator 37

The right hand side of (5.5) is just the Mellin transform of the function $e^{-t\lambda}$. Thus for $Re(s) > \frac{m}{d}$

$$\begin{aligned}\Gamma(s)\zeta_A(s) &= \int_0^\infty dt\, t^{s-1} tr'e^{-tA} \\ &= (\int_0^1 dt + \int_1^\infty dt) t^{s-1} tr'e^{-tA},\end{aligned} \quad (5.6)$$

where $tr'e^{-tA} = \sum' e^{-t\lambda_n}$.

Because of (5.3) the integral over $[1, \infty)$ is bounded by a constant times the integral $\int_1^\infty dt\, t^{s-1}e^{-t\lambda_0}$, λ_0 being the smallest eigenvalue different from zero, so it defines an entire function of s.

For $0 < t < 1$ the semi-group e^{-tA} has a kernel $e^{-tA}(x, y)$ defined on $M \times M$ whose diagonal part $e^{-tA}(x, x)$ admits the following asymptotic expansion:

$$e^{-tA}(x,x) = \sum_{n=0}^{N-1} t^{\frac{n-m}{d}} a_n(x) + O(t^{\frac{N-m}{d}}), t \to 0^+, \quad (5.7)$$

for each $N = 1, 2, ...$, the $a_n(x)$ being certain A-dependent scalar functions. Let us set $\gamma_0 := \dim \ker(A)$ (which is finite), then $tr'e^{-tA} = tre^{-tA} - \gamma_0$ and thus by (5.7) the integral over $[0,1]$ splits into the sum of the following two integrals:

$$\begin{aligned}I_1(s) &:= \mathrm{Vol}_g(M) \int_0^1 dt\, t^{s-1} O(t^{\frac{N-m}{d}}) \\ I_2(s) &:= \int_0^1 dt\, t^{s-1} \left(\int_M \sum_{n<N} t^{\frac{n-m}{d}} a_n(x) d\mu_g(x) - \gamma_0 \right),\end{aligned} \quad (5.8)$$

$d\mu_g$ denoting the volume form on M with respect to the Riemannian metric g.

$I_1(s)$ is a holomorphic function for $Re(s) > \frac{m-N}{d}$ so computing $I_2(s)$ explicitly and setting $a_n := \int_M a_n(x) d\mu_g(x)$ we obtain

$$\Gamma(s)\zeta_A(s) = \sum_{n<N} \frac{a_n}{s - \frac{m}{d} + \frac{n}{d}} - \frac{\gamma_0}{s} + R_N(s) \quad (5.9)$$

with $R_N(s)$ a function which is holomorphic for $Re(s) > \frac{m-N}{d}$. Since N can be arbitrarily chosen, formula (5.9) can be used to extend $\Gamma(s)\zeta_A(s)$ meromorphically to the whole plane. $\Gamma(s)$ being already meromorphic (with poles located at $0, -1, -2, ...$), this defines a meromorphic extension for $\zeta_A(s)$. Dividing the right hand side of (5.9) by $\Gamma(s)$ it is clear that $s = 0$ is a regular point for $\zeta_A(s)$ since $s\Gamma(s) = \Gamma(s+1)$.

I.5: Zeta-function and heat-kernel determinants of an operator

As a consequence of this analysis it makes sense to set

$$\det{}' A := e^{-\zeta'_A(0)}, \tag{5.10}$$

generalizing (5.1). $\det{}'A$ is called the <u>zeta-function determinant</u> of A.

Let us now consider the case when A is a second-order differential operator on a compact boundaryless surface (this is the case relevant for string theory). To be even more explicit we shall choose A to be the negative of the Laplace-Beltrami operator Δ_g, i.e. $A = -\Delta_g$ with Δ_g given by (4.7), the analysis remaining essentially the same in the general case. Fixing therefore $m = d = 2$ and taking $N = 4$ we get from (5.7)

$$e^{t\Delta_g}(x,x) = \frac{a_0(x)}{t} + a_2(x) + O(t), t \to 0^+ \tag{5.11}$$

(for n odd $a_n(x) \equiv 0$; see e.g. [G1,2]).

For the coefficients in (5.11) the values are found to be $a_0(x) = \frac{1}{4\pi}$, $a_2(x) = \frac{1}{24\pi}R_g(x)$ where $R_g(x)$ is the <u>scalar curvature</u> at the point x (e.g. [G1,2], [Cha]). Applying the Gauss-Bonnet theorem and taking into account that $\gamma_0 = 1$ we obtain

$$tr' e^{t\Delta_g} = \frac{\mathrm{Vol}_g(M)}{4\pi t} + \left(\frac{\chi(M)}{6} - 1\right) + O(t), t \to 0^+, \tag{5.12}$$

$\chi(M)$ being the Euler-Poincaré characteristic of M, and from (5.9)

$$\zeta_{-\Delta_g}(s) = \frac{1}{\Gamma(s)}\left\{\frac{\mathrm{Vol}_g(M)}{4\pi(s-1)} + \left(\frac{\chi(M)}{6} - 1\right)\frac{1}{s} + R_4(s)\right\} \tag{5.13}$$

for $Re(s) \to 1$.

To simplify the notation we set $a = \mathrm{Vol}_g(M)/4\pi$ and $b = \frac{\chi(M)}{6} - 1$. Considering the Taylor expansion $\Gamma(s)^{-1} = s + \gamma s^2 + O(s^3), s \to 0, \gamma$ being the Euler constant, it is possible to calculate $\zeta'_{-\Delta_g}(s)$ for small s. The result is

$$\zeta'_{-\Delta_g}(s) = \frac{\gamma s^2 - 2\gamma s - 1}{(s-1)^2}a + \gamma b + H(s) + O(s), s \to 0, \tag{5.14}$$

where $H(s)$ is a certain holomorphic function such that $H(0) = R_4(0)$ (we remark that in (5.14) $a, H(s)$ and $O(s)$ all depend on g). Thus

$$\zeta'_{-\Delta_g}(0) = -a + \gamma b + \int_0^1 \frac{dt}{t}O(t) + \int_1^\infty \frac{dt}{t}tr' e^{t\Delta_g} \tag{5.15}$$

where we have used the explicit expression for $R_4(0)$.

I.5: Zeta-function and heat-kernel determinants of an operator 39

As we shall see, (5.15) allows us to compare the zeta-function determinant with the heat-kernel determinant of an operator, a concept which we now introduce (following [AHKPS2]).

Let A be an operator which satisfies the hypothesis stated at the beginning and let A' denote the restriction of A to the subspace orthogonal to Ker A. For $\varepsilon > 0$ let us consider the function $h_\varepsilon : (0,\infty) \longrightarrow (0,\infty)$ given by

$$h_\varepsilon(\lambda) := \exp(-\int_\varepsilon^{+\infty} \frac{dt}{t} e^{-t\lambda}). \tag{5.16}$$

Then by the spectral theorem the operator $h_\varepsilon(A')$ makes sense as a positive self-adjoint bounded operator. For a certain class of operators which we now describe, $h_\varepsilon(A')$ is of the form "1+ trace class" so that we can define the determinant of $h_\varepsilon(A')$.

Let E be a smooth vector bundle based on a boundaryless C^∞ 2-dimensional real compact manifold with fibers of finite dimension and let $C^\infty(E)$ be the vector space of smooth sections of E. We shall denote by $H \equiv L^2(E)$ the closure of $C^\infty(E)$ w.r.t. the L^2 scalar product induced by an L^2 Riemannian metric $<\cdot,\cdot>$ on E. The set of positive elliptic self-adjoint (w.r.t. $<\cdot,\cdot>$) operators on H will be denoted by $\mathrm{Ell}^+(E)$.

Proposition 5.1: For $\varepsilon > 0$, $A \in \mathrm{Ell}^+(E)$, $h_\varepsilon(A')$ is of the type "1+ trace class".

Proof: From the asymptotic behaviour of the eigenvalues λ_n of A given by $\lambda_n \sim Cn^\alpha$, $\alpha, C > 0$ and the monotonicity of h_ε there follows the existence of a constant $C_1 > 0$ such that $|\log h_\varepsilon|(\lambda_n) \leq |\log h_\varepsilon(C_1 n^\alpha)|$ for n large enough. We observe that $|\log h_\varepsilon(\lambda)| \leq \frac{1}{\varepsilon} e^{-\varepsilon\lambda}$ for $\lambda \geq 1$ and deduce from this the absolute convergence of the series with general term $|\log h_\varepsilon(\lambda_n)|$. In particular we have that $\lim_{n\to\infty} h_\varepsilon(\lambda_n) = 1$. Let us now set $\gamma_n \equiv 1 - h_\varepsilon(\lambda_n)$. Since $|\gamma_n|$ is bounded asymptotically for $n \to \infty$ by $c|\log h_\varepsilon(\lambda_n)|$ (for some constant c), we have $\sum_{n=1}^\infty |\gamma_n| < \infty$ as a result of the absolute convergence proved above. Hence $h_\varepsilon(A')$ is of the form "1+ trace class".

∎

As a consequence of Prop. 5.1 it is possible to use the usual definition (see e.g. [S2])

$$\mathrm{Det}(h_\varepsilon(A')) := \exp(tr(\log h_\varepsilon(A')), \tag{5.17}$$

the right hand side being a finite number. The $\underline{\varepsilon\text{-regularized heat-kernel determinant}}$ of A is then defined by

$$\mathrm{Det}'_\varepsilon(A) := \mathrm{Det}(h_\varepsilon(A')). \tag{5.18}$$

I.5: Zeta-function and heat-kernel determinants of an operator

Sending the regulator ε to zero in (5.18) as it stands, we get infinity. It is necessary therefore to subtract from (5.18) the divergent part before taking such a limit. In physics this procedure is called renormalization, which motivates the name <u>renormalized heat-kernel determinant</u> of A (or shortly <u>heat-kernel determinant</u> of A) for the quantity

$$\operatorname{Det}'(A) := \exp\left\{\lim_{\varepsilon \to 0}\left[\log \operatorname{Det}'_\varepsilon(A) - (\text{divergent terms})\right]\right\}. \tag{5.19}$$

Remark: The notions of zeta-function and heat-kernel determinant can of course be extended to operators acting on vector bundles based on surfaces with a boundary, since when one imposes boundary conditions on the operator, such as Dirichlet or Neumann boundary conditions, its spectrum has the "same qualitative properties" as the spectrum of the same operator on a boundaryless surface. In particular, if it was discrete, it stays discrete and the eigenvalues have the same asymptotic behavior.

It is of some interest to set up the relationship between the zeta-function and heat-kernel determinants of an operator (see also [AP2]). By (5.18) and (5.16) we have

$$\log \operatorname{Det}'_\varepsilon(A) = \int_\varepsilon^{+\infty} \frac{dt}{t} tr' e^{-tA} \tag{5.20}$$

(since $tr e^{-tA'} = tr' e^{-tA}$), therefore in the case $A = -\Delta_g$ we get from (5.15), (5.12) and (5.10)

$$\log \det'(-\Delta_g) = a - \gamma b - \lim_{\varepsilon \to 0} \int_\varepsilon^1 \frac{dt}{t}(tr' e^{t\Delta_g} - \frac{a}{t} - b) - \int_1^\infty \frac{dt}{t} tr' e^{t\Delta_g}$$
$$= -\gamma b + \lim_{\varepsilon \to 0}\left[\log \operatorname{Det}'_\varepsilon(-\Delta_g) - (\frac{a}{\varepsilon} + b \log \varepsilon)\right]. \tag{5.21}$$

From (5.19) and (5.21) we deduce

$$\operatorname{Det}'(-\Delta_g) = e^{\gamma(\chi(M)/6 - 1)} \det'(-\Delta_g) \tag{5.22}$$

(see also [D], [Sch]) which shows that the heat-kernel determinant of $-\Delta_g$ coincides with the zeta-function determinant up to a finite positive constant depending only on the topology of M.

I.6 The Faddeev–Popov procedure

The aim of this section is to give a description of the Faddeev–Popov procedure for gauge field theories first introduced by Faddeev and Popov in the case of Yang–Mills theories [FP]. It yields a device to compute formally functional integrals over configurations of fields, the integrand being invariant under the action of the gauge group. This invariance leads to an infinite expression and the Faddeev–Popov procedure gives a prescription for "factoring out" the infinite volume of the gauge group. More precisely, one considers the integration of a G-invariant function on a principal fiber bundle with structure group G, picking out a well chosen set of representatives (a local section of the bundle) of the orbits on which the integration will actually be done. The Faddeev–Popov procedure then tells us how to relate the integral along this section with the integral on the whole orbit space. The presentation that follows is close to [Pa2], [Pa3]. Further discussions concerning the interpretation of the Faddeev–Popov procedure were presented in [AP2], [AP3].

Going from an integral on the configuration space to an integral on a section gives rise to a Jacobian determinant, the Faddeev–Popov determinant; it is, up to a finite-dimensional determinant which depends on the choice of the slice, entirely determined by the geometric data, namely, the fiber bundle $P \to P/G$ equipped with a Riemannian structure.

We shall assume the fibration is trivial and we shall work with trivial principal bundles (since the procedure we shall describe is a local one we can always restrict ourselves to an open subset of the base space).

In general, the Riemannian structure on the manifold P is not invariant under the whole group G but only under a subgroup H of G. Here we shall assume G can be described as a semidirect product of H with a group K:

$$G = H \odot K.$$

The group G corresponds to the invariance group of the classical action of the theory, whereas the group H arises as the remaining invariance group of the quantized action. In string theory for example, only in critical dimensions is the invariance group of the quantized action the whole group G.

In order to simplify the presentation, we shall first describe the Faddeev–Popov determinant in the case where $G = H$ and then generalize the construction to the case $K \neq \{0\}$ which entails the case considered in string theory.

I.6.1. The Faddeev–Popov map

Let P be a C^∞ trivial principal fiber bundle on a C^∞ manifold V with structure group G equipped with a C^∞ smooth Riemannian structure. Let

I.6: The Faddeev-Popov procedure

$\pi : P \to P/G \cong V$ denote the canonical projection and $\sigma : V \to P$ be a C^∞ section of P. Set $\Sigma \equiv \sigma(V)$. We shall make the following assumptions on the geometric data which are fulfilled in the context of string theory, as we shall see in part II.

Hyp 1: The quotient space P/G is finite dimensional.

Hyp 2: The group G is an infinite-dimensional smooth Fréchet–Lie group equipped with a smooth Riemannian structure, and its tangent space $T_e G$ at the unit point is the space of C^∞ sections of some vector bundle E with finite-dimensional fibers on a compact smooth surface S, with appropriate boundary conditions in case $\partial S \neq \emptyset$. We shall denote by $L^2(E)$ the closure of $C^\infty(E)$ with respect to the scalar product induced by the Riemannian structure.

We shall furthermore make an assumption on the tangent action of the group G on the tangent space to P, which we are about to describe. Let the group G act smoothly on P by the right action

$$G \times P \to P$$
$$(a,p) \to R_a \cdot p = p \cdot a. \tag{6.1}$$

To each $p \in P$, there corresponds a unique element $x = \sigma(\pi(p)) \in \Sigma$ such that $p = R_a x$ for some $a \in G$, and we set

$$W_p \equiv T_p(R_a \Sigma). \tag{6.2}$$

We also introduce the space

$$V_p \equiv T_p(O_p) \tag{6.3}$$

tangent to the fiber

$$O_p \equiv \pi^{-1}(\pi(p)) \tag{6.4}$$

at point p. Then $V_p = \text{Im}(D_a \theta_x)$ where $D_a \theta_x$ is the tangent map at point $a \in G$ to:

$$\theta_x : G \to O_p$$
$$a \to R_a \cdot x = x \cdot a. \tag{6.5}$$

Set $\tau_x = D_e \theta_x$, where e is the identity element of the group G. Then $D_e \theta_x = R_a^* \tau_x D R_a^{-1}$ where DR_a is the tangent map to the right hand side multiplication in G, namely $R_a : G \to G$, $b \to ba$. Let us now assume G is a semi-direct product $G = H \odot K$ of two subgroups H and K and let us denote by τ_x' the tangent map to $\theta_x' : H \to X$ $a' \to R_{a'} x = xa'$ at $e \in H$ and τ_x'' the tangent map to $\theta_x'' : K \to X$, $a'' \to R_{a''} x = xa''$ at $e \in K$. Let $G = H \odot K$ act on X by $\Theta((a', a''), x) = R_{a'}(R_{a''}x)$. Then for $a = (a', a'') \in G$, and $x \in X$, when identifying $T_a G$ with $T_{a'} H \times T_{a''} K$, we have

$$D_a \theta_x = R_{a'}^* \tau_{R_{a''}x}' D R_{a'}^{-1} + R_a^* \tau_x'' D R_{a''}^{-1}.$$

I.6: The Faddeev-Popov procedure

Notice that $\tau_x = \tau'_x + \tau''_x$ where as before $\tau_x = D_e \theta_x$ but that the relation $D_a \theta_x = R_a^* \tau_x D R_a^{-1}$ does not hold with the identification of $T_a G$ with $T_{a'} H \times T_{a''} K$. We shall assume that the operator $D_a \theta_x$ fulfills the following requirement:

Hyp 3: The operator $D_a \theta_x$ is a differential operator on S of order at least 1 with smooth coefficients which is injective and has an injective symbol.

Remark: In particular, the operator $D_a \theta_x$ is densely defined on the closure of $L^2(E)$ and its adjoint operator $(D_a \theta_x)^*$ is well defined. Under the above hypothesis, the operator $(D_a \theta_x)^* D_a \theta_x$ is an elliptic operator on S.

The tangent space to the total space P at point $p = R_a \cdot x, a \in G, x \in \Sigma$ splits up on the one hand into a direct sum

$$T_p P = V_p \oplus T_p(R_a \Sigma) = \mathrm{Im} D_a \theta_x \oplus W_p \qquad (6.6)$$

and on the other hand into an orthogonal sum

$$T_p P = \mathrm{Im} D_a \theta_x \oplus \ker (D_a \theta_x)^* \qquad (6.7)$$

since the symbol of $D_a \theta_x$ is injective (see e.g. [Eb]).

Since the operator $D_a \theta_x$ has an injective symbol, the image space $\mathrm{Im} D_a \theta_x$ is closed and we can define the orthogonal projection π_p on $\mathrm{Im} D_a \theta_x$ with respect to the scalar product $< \cdot, \cdot >_p$ induced on $T_p P$ by the Riemannian structure on P.

Lemma 6.1: The map defined for $p = R_a x$ by:

$$\begin{aligned} F_p : T_e G \times W_p &\to \mathrm{Im} D_a \theta_x \times \ker(D_a \theta_x)^* \\ (u, w) &\to (D_a \theta_x \circ DR_a u + \pi_p w, (1 - \pi_p) w) \end{aligned}$$

is one to one and onto.

Proof: This follows clearly from the splittings (6.6) and (6.7) and the injectivity of $D_a \theta_x$ which is a consequence of the diffeomorphism $G \times \Sigma \simeq P$. ∎

We shall call F_p the Faddeev-Popov map and our aim is to compute its determinant. Seen as an operator from $T_e G \times W_p$ to $\mathrm{Im} D_a \theta_x \times \ker(D_a \theta_x)^*$ the operator F_p can be represented by a matrix

$$F_p = \begin{bmatrix} D_a \theta_x \circ DR_a & \pi_p \\ 0 & \mathbb{1} - \pi_p \end{bmatrix}. \qquad (6.8)$$

Setting

$$T_p \equiv \begin{bmatrix} D_a \theta_x \circ DR_a & 0 \\ 0 & \mathbb{1} \end{bmatrix} \qquad (6.9\,a)$$

(which should not be confused with a tangent space at p!) and

$$R_p \equiv \begin{bmatrix} 0 & \pi_p \\ 0 & -\pi_p \end{bmatrix} \qquad (6.9\ b)$$

we can write
$$F_p = T_p + R_p.$$

Under the assumption Hyp 1, the operator R_p is clearly a finite-rank operator since it acts on W_p, which is finite dimensional. This operator will give rise to a finite-dimensional determinant which depends on the choice of the slice Σ. However, the operator T_p is built up from the elliptic operator $D_a\theta_x$, for which we shall define a determinant following the lines of section 5.

I.6.2. The Faddeev–Popov determinant: the case $G=H$

In this section, we assume that $G = H$ so that $D_a\theta_x = R_a^*\tau_x \circ DR_a^{-1}$, where $\tau_x = D_e\theta_x : T_eG \to T_x\Sigma$. We remark that in this case, the assumption Hyp 3 can be replaced by the same assumption on the operator τ_x, namely:

Hyp 3′: The operator τ_x is a differential operator on S of order 1 with smooth coefficients which is injective and has an injective symbol.

The Riemannian metric on P being invariant under the action of G, it is equivalent to compute the determinant of $F_p = \begin{bmatrix} D_a\theta_x \circ DR_a & \pi_p \\ 0 & \mathbb{1} - \pi_p \end{bmatrix}$ with $p = R_a x$ or of $F_x = \begin{bmatrix} \tau_x & \pi_x \\ 0 & \mathbb{1} - \pi_x \end{bmatrix}$ since $F_p = R_a^* F_x$.

The operator F_x is densely defined on $L^2(E) \times W_x$ and hence has a well-defined adjoint operator F_x^*; since F_x takes C^∞ sections into C^∞ sections and since it is built up from differential operators with smooth coefficients, the operator F_x^* is defined on the image of F_x and the operator $F_x^* F_x$ is well defined on $L^2(E) \times W_x$. It can therefore be extended to a self-adjoint operator which we shall denote by the same symbol. Since T_x is invertible, we have
$$F_x^* F_x = (\mathbb{1} + T_x^{-1} R_x)^* (T_x^* T_x)(\mathbb{1} + T_x^{-1} R_x).$$

In general, since $T_x^* T_x$ does not lie in a class for which the determinant is defined, neither does $F_x^* F_x$, and we need to regularize it in order to define a regularized determinant along the lines of section 5. Since the operator $\tau_x^* \tau_x$ is elliptic (Hyp 3′), applying the results of the proposition in section 5, we can define for $\varepsilon > 0$, the operator $(\tau_x^* \tau_x)_\varepsilon \equiv h_\varepsilon(\tau_x^* \tau_x)$ which is of the form "$\mathbb{1}+$ trace class operator". Let us set for $\varepsilon > 0$

$$(F_x^* F_x)_\varepsilon \equiv (\mathbb{1} + T_x^{-1} R_x)^* (T_x^* T_x)_\varepsilon (\mathbb{1} + T_x^{-1} R_x), \qquad (6.10)$$

whereby
$$(T_x^* T_x)_\epsilon \equiv \begin{bmatrix} (\tau_x^* \tau_x)_\epsilon & 0 \\ 0 & \mathbb{1} \end{bmatrix}. \tag{6.11}$$

Proposition 6.1: Under assumptions Hyp 1, Hyp 2 and Hyp 3' and for $\epsilon > 0$, the operator $(F_x^* F_x)_\epsilon$ is an injective operator of the form "$\mathbb{1}$ + trace class operator" and therefore has a well defined determinant given by

$$\mathrm{Det}(F_x^* F_x)_\epsilon = \mathrm{Det}_\epsilon(\tau_x^* \tau_x) \cdot \frac{\det^2(<\Psi_x^i, \xi_x^j>_x)}{\det(<\Psi_x^i, \Psi_x^j>_x)}$$

where $\Psi_x^i, i = 1, \cdots, \dim P/G$ is a basis of $\mathrm{Ker}(\tau_x^*)$ and $\xi_x^i, i = 1, \cdots, \dim P/G$ is an orthonormal basis of W_x.

Proof: The fact that the operator $(F_x^* F_x)_\epsilon$ is of the form "$\mathbb{1}$ + trace class" follows from the fact that the operator R_x being of finite rank (Hyp 1), $(F_x^* F_x)_\epsilon$ is a product of three operators of the form "$\mathbb{1}$+ trace class" (using Hyp 3') and hence also is of that form. Furthermore, its determinant is the product of the determinants of the three operators. It is clear that $\mathrm{Det}(T_x^* T_x)_\epsilon = \mathrm{Det}_\epsilon(\tau_x^* \tau_x)$ by formula (6.11). On the other hand,

$$\mathrm{Det}(\mathbb{1} + T_x^{-1} R_x) = \mathrm{Det}((\mathbb{1} - \pi_x) \restriction W_x) \tag{6.12}$$

(with $\restriction W_x$ denoting restriction to W_x), since $T_x^{-1} R_x$ is the sum of a nilpotent matrix $\begin{bmatrix} 0 & \tau_x^{-1} \pi_x \\ 0 & 0 \end{bmatrix}$ (with operator elements) with the matrix $\begin{bmatrix} \mathbb{1} & 0 \\ 0 & \mathbb{1} - \pi_x \end{bmatrix}$. An easy computation then yields the expression of the determinant in (6.12):

$$\det((\mathbb{1} - \pi_x) W_x) = \det(<\tilde{\Psi}_x^i, \xi_x^j>_x))$$

with the notation of the proposition and where $\tilde{\Psi}_x^i, i = 1, \cdots, \dim P/G$ is an orthonormal basis of W_x. Going to a not necessarily orthonormal basis $\Psi_x^i, i = 1, \cdots, \dim P/G$ yields the expression given in the proposition. ∎

In conformity with section 5 we can introduce the following definition:

Definition 6.1: The heat-kernel renormalized Faddeev–Popov determinant is defined for $p = R_a x, a \in G, x \in \Sigma$ by

$$\mathrm{Det} F_p = \mathrm{Det} F_x = \left(\lim_{\epsilon \to 0} [\exp(\log \mathrm{Det}(F_x^* F_x)_\epsilon - \text{divergent terms})] \right)^{\frac{1}{2}}.$$

From the above consideration we have, for $p = R_a x, a \in G, x \in \Sigma$,

$$\mathrm{Det}(F_p) = \mathrm{Det}(F_x) = \mathrm{Det}(\tau_x) \cdot \frac{\det(<\Psi_x^i, \xi_x^j>_x)}{\det(<\Psi_x^i, \Psi_x^j>_x^{\frac{1}{2}})}$$

I.6.3. The Faddeev–Popov determinant: the general case

We now drop the hypothesis that $G = H$. The group G is here a semidirect product of two groups H and K where H acts on K by an action satisfying the condition

$$(k^{h_1})^{h_2} = k^{h_1 h_2} \quad \forall h_1, h_2 \in H, \quad \forall k \in K.$$

The group H acts on P through isometries, but not on K. We assume $D_a \theta_x = R_{a'}^* \tau_{R_{a''} x} \circ DR_a$ where $a = (a', a'') \in G = H \cdot K$. Since the Riemannian metric on P is H invariant, the determinant of $F_p = R_{a''} F_{\bar{x}}$ will coincide with that of $F_{\bar{x}}$, where $\bar{x} = R_{a'} x$. Here again, we therefore only need to compute the determinant of an operator of the form $F_{\bar{x}} = \begin{bmatrix} \tau_{\bar{x}} & \pi_{\bar{x}} \\ 0 & \mathbb{1} - \pi_{\bar{x}} \end{bmatrix}$
where $\bar{x} = R_{a''} x$, $x \in E$, $a'' \in K$.

Let us denote by $\tau_{\bar{x}}^H$ (resp. $\tau_{\bar{x}}^K$) the restriction of $\tau_{\bar{x}}$ to $T_e H$ (resp. $T_{a''} K$). The operator $\tau_{\bar{x}}$ can be seen as an operator from $T_e H \times T_{a''} K$ to $(\mathrm{Im}\tau_{\bar{x}}^K)^\perp \times \mathrm{Im}\tau_{\bar{x}}^K$ and represented by the matrix

$$\tau_{\bar{x}} = \begin{bmatrix} A_{\bar{x}} & 0 \\ C_{\bar{x}} & B_{\bar{x}} \end{bmatrix} \tag{6.13}$$

where

$$A_{\bar{x}} = (\mathbb{1} - \pi_{\bar{x}}^K)\tau_{\bar{x}}^H, \quad B_{\bar{x}} = \tau_{\bar{x}}^K, \quad C_{\bar{x}} = \pi_{\bar{x}}^K \tau_{\bar{x}}^H$$

and where $\pi_{\bar{x}}^K$ is the orthogonal projection onto $\mathrm{Im}\tau_{\bar{x}}^K$ which is closed since $\tau_{\bar{x}}$ has an injective symbol.

In the case of interest to us, namely string theory (see chapter II), the operator $B_{\bar{x}}$ is an isometry $i_{\bar{x}}$ from $T_{a''} K$ into $T_{\bar{x}} P$ and we shall write $B_{\bar{x}} = \mathbb{1}$, identifying $T_{a''} K$ with $i_{\bar{x}}(T_{a''} K)$, since this has no influence on the level of determinants.

The assumption on the injectivity of the symbol of $\tau_{\bar{x}}$ which follows from that of $D_a \theta_x$, namely Hyp 3, yields that the operator $A_{\bar{x}}$ has an injective symbol and hence that the operator $A_{\bar{x}}^* A_{\bar{x}}$ is elliptic on S. Applying the results of section 5, we can therefore define for $\varepsilon > 0$ the operator $h_\varepsilon(A_{\bar{x}}^* A_{\bar{x}}')$ (note that $A_{\bar{x}}$ is injective whenever $\tau_{\bar{x}}$ is injective).

We now proceed to generalize the results of the above paragraph to this general case.

Lemma 6.2: For $\varepsilon > 0$ the operator

$$(\tau_{\bar{x}}^* \tau_{\bar{x}})_\varepsilon \equiv \begin{bmatrix} \mathbb{1} & (\mathbb{1} + \varepsilon C_{\bar{x}}^* C_{\bar{x}})^{-2} C_{\bar{x}}^* 0 & \mathbb{1} \end{bmatrix} \begin{bmatrix} h_\varepsilon(A_{\bar{x}}^* A_{\bar{x}}) & 0 \\ 0 & \mathbb{1} \end{bmatrix}$$

$$\begin{bmatrix} \mathbb{1} & 0 \\ C_{\bar{x}}(\mathbb{1} + \varepsilon C_{\bar{x}}^* C_{\bar{x}})^{-2} & \mathbb{1} \end{bmatrix}$$

I.6: The Faddeev-Popov procedure

is an injective operator of the form "$\mathbb{1}+$ trace class operator" and we have

$$\mathrm{Det}(\tau_{\bar{x}}^*\tau_{\bar{x}})_\varepsilon = \mathrm{Det}_\varepsilon(A_{\bar{x}}^*A_{\bar{x}}).$$

Proof: The operator $(\tau_{\bar{x}}^*\tau_{\bar{x}})_\varepsilon$ is of the form "Identity + trace class" as a product of three operators of this form. The middle matrix clearly represents an operator of this form since $h_\varepsilon(A_{\bar{x}}^*A_{\bar{x}}) - \mathbb{1}$ is trace class by the results of section 5. As for the two extreme operators, they are also of this form since the operator $C_{\bar{x}}(\mathbb{1}+\varepsilon C_{\bar{x}}^*C_{\bar{x}})^{-2}$ and its adjoint are inverses of elliptic operators of degree at least 2.

The determinant of $(\tau_{\bar{x}}^*\tau_{\bar{x}})_\varepsilon$ is the product of the determinants of these three operators. The operators on either end are of the form "Identity + nilpotent operator" and therefore have a determinant equal to 1, and the middle operator has determinant $\mathrm{Det}_\varepsilon(A_{\bar{x}}^*A_{\bar{x}})$, which yields the expression given in Lemma 6.2.

■

We can now extend the definition of the Faddeev–Popov determinant to this general case. First we consider its regularized version.

Proposition 6.2: For $\bar{x} = R_{a''}x, a'' \in K, x \in \Sigma$, $p = R_a x, a = (a', a'') \in HK$, define for $\varepsilon > 0$ the ε-cutoff operator $(F_{\bar{x}}^*F_{\bar{x}})_\varepsilon$ as in formula (6.10), replacing x by $\bar{x} = R_{a''}x$, and where now $(T_{\bar{x}}^*T_{\bar{x}})_\varepsilon$ is given by formula (6.11) and $(\tau_{\bar{x}}^*\tau_{\bar{x}})_\varepsilon$ by the formula in Lemma 6.2. Set $(F_p^*F_p)_\varepsilon \equiv R_{a'}^*(F_{\bar{x}}^*F_{\bar{x}})$. This operator is an injective operator of the form "Identity + trace class" and its determinant is given by the formula

$$\mathrm{Det}(F_p^*F_p)_\varepsilon = \mathrm{Det}_\varepsilon(A_{\bar{x}}^*A_{\bar{x}}) \left[\frac{\det^2(<\Psi_{\bar{x}}^i, \zeta_{\bar{x}}^j>_p)}{\det(<\Psi_{\bar{x}}^i, \Psi_{\bar{x}}^j>_p)}\right]$$

with the notation of Proposition 6.1, replacing x by $\bar{x}$.

Proof: The proof goes as in Proposition 6.1 using the results of Lemma 6.2

■

We can now define the heat-kernel renormalized Faddeev–Popov determinant $\mathrm{Det}(F_p)$ by the same formula as in Def. 6.1. We obtain, for $p = R_{a'}\bar{x}, a' \in H, \bar{x} = R_{a''}x, a'' \in K, x \in \Sigma$,

$$\mathrm{Det}(F_p) = \mathrm{Det}(A_{\bar{x}}) \cdot \frac{\det(<\Psi_{\bar{x}}^i, \zeta_{\bar{x}}^j>_{\bar{x}})}{\det(<\Psi_{\bar{x}}^i, \Psi_{\bar{x}}^j>_{\bar{x}})^{\frac{1}{2}}}.$$

We remark that Proposition 6.1 arises as a particular case of Proposition 6.2 since when $H = G$, we have $\pi_p^K = 0$, $\bar{x} = x$ and $A_{\bar{x}} = \tau_{\bar{x}}^H = \tau_{\bar{x}} = \tau_x$ so that the Faddeev–Popov determinants then coincide.

I.7 Determinant bundles

In this section we show that the notion of regularized determinants for elliptic operators arises naturally when equipping determinant bundles with a Hermitian metric. As a motivation for the definition of determinant bundle, we first discuss the finite-dimensional case.
Let
$$L : V \to W$$
be a linear map between finite-dimensional vector spaces V and W of the same dimension m equipped with inner products. Putting
$$\det V := \wedge^m(V),$$
we get an induced map
$$\det L : \det V \to \det W,$$
the determinant of L. Unless $V = W$, $\det L$ cannot be interpreted as a number, but rather as an element of $(\det V)^* \otimes \det W$. Its transformation behavior is given w.r.t. bases $e_1, \ldots, e_m$ of V and $f_1, \ldots, f_m$ of W by
$$\det L(e_1 \wedge \ldots \wedge e_m) = Le_1 \wedge \ldots \wedge Le_m =: \Delta_L f_1 \wedge \ldots \wedge f_m. \quad (7.1)$$
We also note that the exact sequence
$$O \to \ker L \to V \xrightarrow{L} W \to \operatorname{coker} L \to O$$
and the multiplicative properties of det allow the identification
$$(\det V)^* \otimes \det W \cong (\det \ker L)^* \otimes (\det \operatorname{coker} L); \quad (7.2)$$
(we use e.g. the inner product on W to identify $\operatorname{coker} L$ with the orthogonal complement of $L(V)$) (for more details about det see e.g. [BeGV]).

In case $\ker L = \{O\}$, the right hand side of (7.2) is identified with the field over which V and W are defined; in our applications this field will be $\mathbb{C}$. If $\ker L = \{O\}$, under the isomorphism (7.2), $\det L$ is identified with $1 \in \mathbb{C}$. If $\ker L \neq \{O\}$, then of course $\det L = O$.

Forming $\det L$ for every $L \in \operatorname{Hom}(V, W) = V^* \otimes W$, we obtain a section det of the trivial line bundle over $V^* \otimes W$ with fiber $(\det V)^* \otimes \det W$.

We now want to generalize these constructions to an infinite-dimensional setting. We let Y be a manifold parametrizing a family $\{D_y\}_{y \in Y}$ of elliptic operators
$$D_y : \Gamma(E_y^1) \to \Gamma(E_y^2),$$

I.7: Determinant bundles

where E_y^1, E_y^2 are Hermitian bundles and $\Gamma(E)$ are the sections of E. We have the following examples in mind.

(1) In the notation of §I.3,§I.4, we consider, for a surface S,

$$Z = \mathcal{M} \times S/\mathcal{D}_o$$
$$\downarrow$$
$$Y = \mathcal{M}/\mathcal{D}_o$$

Since the metric along the fibers of $\mathcal{M} \times S \to \mathcal{M}$ is preserved by the action of $\mathcal{D}_o$, Z also carries a metric along the fibers, and we obtain a projection $\pi : TZ \to T_{\text{vert}}Z$, where the latter is the tangent space along the fibers. One can also attempt to divide by $\mathcal{D}$ instead of $\mathcal{D}_o$ in which case we no longer have manifolds.

If one looks at Dirac operators, one can divide only by a subgroup $\mathcal{D}'$ of finite index in $\mathcal{D}$ preserving a spin structure. For each $y \in Y$, one then has corresponding $\bar{\partial}$-operators, yielding families $\{D_y\}$.

(2) Let $Y = T_p$, $p \geq 2$, Z = universal Teichmüller curve, i.e. the fiber bundle over T_p where the fiber over $y \in T_p$ is the Riemann surface determined by y. Again, one has families of $\bar{\partial}$-operators. One wants to divide by the action of the mapping class group.

(3) σ-models: Let Σ be a Riemann surface, M a manifold, E a Hermitian vector bundle over M with unitary connection ∇. We put

$$Y := C^\infty(\Sigma, M)$$
$$Z := Y \times \Sigma. \tag{7.3}$$

For $\phi \in Y$, we obtain the bundle ϕ^*E with connection $\phi^*\nabla$, and $\bar{\partial}$-operators on Σ couple to this connection to give Dirac operators on ϕ^*E-valued spinor fields.

Since the operators D_y are elliptic, they are Fredholm operators, i.e. have finite-dimensional kernels and cokernels. Thus, each D_y defines a determinant line

$$(\det \ker D_y)^* \otimes (\det \operatorname{coker} D_y).$$

We want to patch these lines together to form a line bundle over Y with a smooth metric and connection.

We define, omitting the subscript y occasionally,

$$\Delta^1 := D^*D : \Gamma(E^1) \to \Gamma(E^1)$$
$$\Delta^2 := DD^* : \Gamma(E^2) \to \Gamma(E^2).$$

Appropriate boundary conditions being assumed, these are nonnegative self-adjoint operators.

Now for each y, D_y defines an isomorphism

$$D_y : (\ker D_y)^\perp \to (\ker D_y^*)^\perp$$
$$\phi \to D_y \phi.$$

If ϕ_n is an eigensection of Δ_y^1 with eigenvalue $\lambda_n > 0$, then $\psi_n = D_y \phi_n$ is an eigensection of Δ_y^2 with the same eigenvalue. Thus, only the multiplicity of the eigenvalue 0 may differ between Δ^1 and Δ^2, and this multiplicity is the index of D:

$$I(D) := \dim \ker D - \dim \operatorname{coker} D.$$

Thus $\Gamma(E_y^1)$ and $\Gamma(E_y^2)$ differ by the kernels of D_y and D_y^*.

If y varies, then $\dim \ker D_y$ and $\dim \operatorname{coker} D_y$ may vary individually (in our applications later on, they will be constant, however), but their difference $I(D_y)$ remains constant (assuming smoothness of D_y in y).

For $\lambda > 0$ not in the spectrum of $\Delta_{y_0}^1$, we let

$$U_\lambda := \{y \in Y : \lambda \text{ is not in the spectrum of } \Delta_y^1\}.$$

U_λ is then an open neighborhood of y_0, and the number of eigenvalues less than λ is constant in each connected component of U_λ.

We let $E_{y,\lambda}^1, E_{y,\lambda}^2$ be the spaces spanned by eigensections of Δ_y^1, Δ_y^2 respectively, with eigenvalues less than λ. These are finite-dimensional spaces and over U_λ we have the lines

$$L_{y,\lambda} := (\det E_{y,\lambda}^1)^* \otimes \det E_{y,\lambda}^2. \tag{7.4}$$

Again, we have a canonical isomorphism

$$L_{y,\lambda} \simeq (\det \ker D_y)^* \otimes (\det \ker D_y^*), \tag{7.5}$$

coming from the exact sequence

$$0 \to \ker \Delta_y^1 \to E_{y,\lambda}^1 \xrightarrow{D} E_{y,\lambda}^2 \to \ker \Delta_y^2 \to 0. \tag{7.6}$$

Likewise, we let the symbols $E_{y,(\lambda,\mu)}^1$ etc. correspond to eigenvalues larger than λ, but less than μ. Then over $U_\lambda \cap U_\mu$

$$L_{y,\mu} = L_{y,\lambda} \otimes L_{y,(\lambda,\mu)}, \tag{7.7}$$

with

$$L_{y,(\lambda,\mu)} := (\det E_{y,(\lambda,\mu)}^1)^* \otimes \det E_{y,(\lambda,\mu)}^2. \tag{7.8}$$

I.7: Determinant bundles

As remarked above, D_y, $y \in U_\lambda \cap U_\mu$, induces an isomorphism

$$D_{y,(\lambda,\mu)} : E^1_{y,(\lambda,\mu)} \to E^2_{y,(\lambda,\mu)}, \qquad (7.9)$$

hence also an isomorphism

$$\det D_{y,(\lambda,\mu)} : \det E^1_{y,(\lambda,\mu)} \to \det E^2_{y,(\lambda,\mu)}, \qquad (7.10)$$

and we consider this as a nonvanishing section of $L_{\lambda,\mu}$. We obtain an induced canonical isomorphism

$$\begin{aligned} L_\lambda &\to L_\lambda \otimes L_{\lambda,\mu} = L_\mu \\ S &\to S \otimes \det D_{(\lambda,\mu)} \end{aligned} \qquad (7.11)$$

over $U_\lambda \cap U_\mu$. We can use this isomorphism to patch the L_λ together to obtain a differentiable line bundle

$$L \to Y. \qquad (7.12)$$

The fiber of L over y is isomorphic to

$$\det D_y = (\det \ker D_y)^* \otimes (\det \ker D_y^*), \qquad (7.13)$$

cf. (7.5).
Let us assume now
$$I(D) = 0 \qquad (7.14)$$

(this can easily be achieved by restricting our operators to $(\ker D)^\perp$ and $(\ker D^*)^\perp$ provided their dimensions are constant in y). Moreover $\dim E^1_\lambda = \dim E^2_\lambda$ for all λ. Then each L_λ has a canonical section

$$\det D_\lambda : \det E^1_\lambda \to \det E^2_\lambda, \qquad (7.15)$$

and $\det D_\lambda$ and $\det D_\mu$ correspond under the isomorphism (7.11). Consequently, we obtain a global section

$$\det D$$

of $L \to Y$, and
$$\det D_y = 0 \Leftrightarrow D_y \text{ is not invertible}.$$

(If $I(D) \neq 0$, we define $\det D = 0$).

We now want to define a Hermitian metric on the line bundle $L \to Y$. Each fiber has an induced metric from the metrics on $\Gamma(E^1_{y,\lambda}), \Gamma(E^2_{y,\lambda})$. If

$\phi_1, \ldots, \phi_{k(\lambda)}$ and $\psi_1, \ldots, \psi_{l(\lambda)}$ are bases of $E^1_{y,\lambda}$ and $E^2_{y,\lambda}$, respectively, a section s_λ of L_λ is given by

$$s_\lambda(\phi_1 \wedge \ldots \wedge \phi_{k(\lambda)}) = \sigma_\lambda \, \psi_1 \wedge \ldots \wedge \psi_{l(\lambda)}, \tag{7.16}$$

and its induced norm is given by

$$\|s_\lambda\|^2_{\text{ind}} = |\sigma_\lambda|^2 \, \frac{\det(\langle \psi_\alpha, \psi_\beta \rangle)}{\det(\langle \phi_\mu, \phi_\nu \rangle)}. \tag{7.17}$$

On $U_\mu \cap U_\lambda$, the same section also gives a section s_μ of L_μ via

$$s_\mu(\phi_1 \wedge \ldots \wedge \phi_{k(\lambda)}) = \sigma_\lambda \, \psi_1 \wedge \ldots \wedge \psi_{l(\lambda)} \otimes (D\phi_{k(\lambda)+1} \wedge \ldots \wedge D\phi_{k(\mu)}) \tag{7.18}$$

i.e.

$$s_\mu = s_\lambda \otimes \det D_{(\lambda,\mu)}, \tag{7.19}$$

and

$$\begin{aligned}\|s_\mu\|^2_{\text{ind}} &= \|s_\lambda\|^2_{\text{ind}} \cdot \|\det D_{(\lambda,\mu)}\|^2 \\ &= \|s_\lambda\|^2_{\text{ind}} \prod_{\lambda < \lambda_n < \mu} \lambda_n, \end{aligned} \tag{7.20}$$

where λ_n are the eigenvalues of Δ^1 counted with multiplicity. In order to have the metrics coincide on L_λ and L_μ, we thus have to define

$$\|s_\lambda\|^2_Q = \|s_\lambda\|^2_{\text{ind}} \cdot (\det D^*D \,|_{\lambda_n > \lambda}) \,^{\dagger)} \tag{7.21}$$

where

$$\det D^*D \,|_{\lambda_n > \lambda} = \prod_{\lambda_n > \lambda} \lambda_n$$

is defined via zeta-function regularization, namely

$$\zeta_\lambda(s) := \sum_{\lambda_n > \lambda} \frac{1}{\lambda_n^s} = \text{tr}\left((D^*D \,|_{\lambda_n > \lambda})^{-s}\right). \tag{7.22}$$

We have shown in section I.5 that, as a consequence of the Weyl estimates for the asymptotic growth of the eigenvalues of elliptic operators, $\zeta_\lambda(s)$ is holomorphic for
Re $s > \frac{1}{2} \dim$ (base of E_y) and admits a meromorphic continuation to $\mathbb{C}$ which is holomorphic at $s = 0$. Following the previous discussions we define therefore, as in (5.10),

$$\det D^*D \,|_{\lambda_n > \lambda} := \exp(-\zeta'_\lambda(0)). \tag{7.23}$$

$^{\dagger)}$ The subscript Q refers to Quillen.

I.7: Determinant bundles

The important property of the zeta-function regularization is

$$\det D^*D \mid_{\lambda_n > \lambda} = (\det D^*D \mid_{\lambda_n > \mu}) \cdot \prod_{\lambda < \lambda_n < \mu} \lambda_n \qquad (7.24)$$

for $\mu > \lambda$ on $U_\lambda \cap U_\mu$. Therefore, the definition of $\|s_\lambda\|_Q^2$ in (7.21) is independent of the choice of λ. In other words, if $\phi_1, \ldots, \phi_k$ is a basis for $\ker D_y$, and $\psi_1, \ldots, \psi_l$ a basis for $\ker D_y^*$, and if the section s is given by

$$s(\phi_1, \wedge \ldots \wedge \phi_k) = \sigma \psi_1, \wedge \ldots \wedge \psi_l,$$

then

$$\|s(y)\|_Q^2 = |\sigma(y)|^2 \frac{\det \langle \psi_\alpha, \psi_\beta \rangle}{\det \langle \phi_\mu, \phi_\nu \rangle} \det D_y^* D_y \mid_{\lambda_n > 0}. \qquad (7.25)$$

In a similar way, we can construct a unitary connection ∇ on $L \to Y$. When D is invertible, it is given via

$$\nabla \det D := \operatorname{tr}(\tilde{\nabla} D \cdot D^{-1}) \det D. \qquad (7.26)$$

Here $\tilde{\nabla}$ is the connection induced by the connection on E and the connection on the tangent spaces to the fibers of Z, but with a correction term to compensate for the changing volumes of the fibers. Again, the right hand side of (7.26) is defined via zeta-function regularization ($\det' = \det D$ since D is invertible) from

$$\omega(s) = \operatorname{tr}((DD^*)^{-s} \tilde{\nabla} D \cdot D^{-1}).$$

For details, we refer to [BF], [F1], [F2].

Before we apply the above definitions to the operators $\bar{\partial}^* \bar{\partial}$ and $\bar{\partial} \bar{\partial}^*$, let us examine the kernels of the $\bar{\partial}$ and $\bar{\partial}^*$ operators.

For later purposes, it will be convenient to introduce a more abstract terminology. In particular, it will be important to reflect the duality obtained by (4.2) more carefully in our notation. Thus untangling the identifications made on the base of (4.2), we write the Cauchy–Riemann operator $\bar{\partial}_n$ (defined in (4.5)) as:

$$\begin{aligned}\bar{\partial}_n : K^n &\to K^n \otimes \bar{K} \\ \phi dz^n &\to \phi_{\bar{z}} dz^n \otimes d\bar{z}.\end{aligned} \qquad (7.27)$$

The kernel of $\bar{\partial}_n$ is formed by holomorphic n-differentials. The sheaf of sections of the line bundle defined by holomorphic 1-differentials which are real on $\partial \Sigma$ is denoted by Ω,

$$\ker \bar{\partial}_n = H^0(\Sigma, \Omega^n). \qquad (7.28)$$

(Often, we shall simply identify Ω and K.)
For this, we note that a holomorphic differential which is real on $\partial\Sigma$, i.e.

$$(u + iv)dz^n,$$

with $v = 0$ on $\partial\Sigma$, locally given as $y = 0$ ($dz = dx + idy$), because of the Cauchy–Riemann equations automatically satisfies

$$u_y = 0 \text{ on } \partial\Sigma \tag{7.29}$$

and thus automatically satisfies the second condition in (4.24) or (4.16), respectively. Actually, and this will be important for our later considerations, a holomorphic differential which is real on $\partial\Sigma$ can be reflected to a holomorphic differential on the Schottky double of Σ.

For this reason, we consider for the moment a compact Riemann surface Σ' without boundary – later interpreted as the Schottky double of Σ. The notation $K, \Omega, \bar{\partial}$ etc. will now refer to Σ'.

We first state some generalities. We let L be a line bundle over Σ'. The Serre duality theorem yields the following relation between cohomology groups:

$$H^1(\Sigma', L) = H^0(\Sigma', \Omega \otimes L^{-1})^*, \tag{7.30}$$

the star denoting the dual vector space. Actually, this is nothing but an abstract formulation of (4.2).

The Riemann–Roch theorem says

$$\dim_{\mathbb{C}} H^0(\Sigma', L) - \dim_{\mathbb{C}} H^1(\Sigma', L) = \deg L - q + 1, \tag{7.31}$$

$q = $ genus Σ'. Hence, with Serre duality (7.29)

$$\dim_{\mathbb{C}} H^0(\Sigma', L) - \dim_{\mathbb{C}} H^0(\Sigma', \Omega \otimes L^{-1}) = \deg L - q + 1. \tag{7.32}$$

Since for a trivial line bundle L, $\dim_{\mathbb{C}} H^0(\Sigma', L) = 1$, as the only global holomorphic sections are given by constant functions, we conclude

$$\dim_{\mathbb{C}} H^1(\Sigma', L) = q. \tag{7.33}$$

From this and the Riemann–Roch theorem for $L = \Omega$ we get

$$\deg \Omega = 2q - 2. \tag{7.34}$$

Looking at our operators $\bar{\partial}_n, \bar{\partial}_n^*$, we note from the above considerations that in the present terminology

$$\ker \bar{\partial}_n = H^0(\Sigma', \Omega^n), \tag{7.35}$$

cf. (7.28), and
$$\operatorname{coker}\bar{\partial}_n = \ker \bar{\partial}_n^* = H^1(\Sigma', \Omega^n) = H^0(\Sigma', \Omega^{1-n})^* \qquad (7.36)$$
by (7.30).
Thus, by the Riemann-Roch theorem
$$\begin{aligned}\operatorname{Ind}\bar{\partial}_n &= \dim_{\mathbb{C}} \ker \bar{\partial}_n - \dim_{\mathbb{C}} \operatorname{coker}\bar{\partial}_n \\ &= \dim_{\mathbb{C}} H^0(\Sigma', \Omega^n) - \dim_{\mathbb{C}} H^0(\Sigma', \Omega^{1-n}) \\ &= n \deg \Omega - q + 1 \\ &= (2n-1)(q-1).\end{aligned} \qquad (7.37)$$

On the other hand, the Kodaira vanishing theorem implies
$$\dim_{\mathbb{C}} H^0(\Sigma', L) = 0 \text{ if } \deg L < 0, \qquad (7.38)$$
hence for $n > 1$,
$$\dim_{\mathbb{C}} H^0(\Sigma', \Omega^{1-n}) = 0. \qquad (7.39)$$
Thus, for $n > 1$
$$\dim_{\mathbb{C}} \ker \bar{\partial}_n = (2n-1)(q-1). \qquad (7.40)$$
For our operators above, we have, cf. (4.10), (7.35),
$$\ker P_g = \ker \bar{\partial}_{-1} = H^0(\Sigma', \Omega^{-1}), \qquad (7.41)$$
and thus by (7.38), (7.34)
$$\ker P_g = 0 \text{ if } q \geq 2. \qquad (7.42)$$
Likewise, by (7.36)
$$\operatorname{coker} P_g = \operatorname{coker} \bar{\partial}_{-1} = H^0(\Sigma', \Omega^2)^* = (\ker \partial_2)^*. \qquad (7.43)$$

Thus, the cokernel of P_g is the dual of the space of holomorphic quadratic differentials. This duality will be important below in our computations of Quillen metrics.

Finally
$$\ker \bar{\partial}_0 = H^0(\Sigma', \Omega^0) = \mathbb{C} \qquad (7.44)$$
$$\operatorname{coker}\bar{\partial}_0 = \ker \bar{\partial}_0^* = H^0(\Sigma', \Omega^1)^* = (\ker \partial_1)^*, \qquad (7.45)$$

i.e. the cokernel of $\bar{\partial}_0$ is the dual of the space of holomorphic (1,0) forms.
We now return to our Riemann surface Σ with boundary and let Σ' be its Schottky double. Thus, there exists an anticonformal involution
$$i : \Sigma' \to \Sigma'$$

with fixed point set $\partial\Sigma$, and

$$\Sigma' = \Sigma \cup i(\Sigma). \tag{7.46}$$

If $\phi \in H^0(\Sigma', \Omega^n)$, then also

$$\bar{\phi} \circ i \in H^0(\Sigma', \Omega^n). \tag{7.47}$$

For $\phi \in H^0(\Sigma', \Omega^n)$, we thus put

$$\phi_1(z) := \phi(z) + \bar{\phi}(i(z)) \tag{7.48}$$
$$\phi_2(z) := \phi(z) - \bar{\phi}(i(z)). \tag{7.49}$$

Then ϕ_1 is invariant under the reflection $\phi \to \bar{\phi} \circ i$, and is real on $\partial\Sigma$. Conversely, each holomorphic differential on Σ which is real on $\partial\Sigma$ can be reflected to such a holomorphic differential on Σ'. This allows us to conclude

$$\dim_{\mathbb{R}} H^0(\Sigma, \Omega^n) = \dim_{\mathbb{C}} H^0(\Sigma', \Omega^n), \tag{7.50}$$

where in the definition of $H^0(\Sigma, \Omega^n)$ we require that elements are real on $\partial\Sigma$, and to make the identification

$$H^0(\Sigma', \Omega^n) = H^0(\Sigma, \Omega^n) \otimes_{\mathbb{R}} \mathbb{C}, \tag{7.51}$$

simply by multiplying elements that are real on $\partial\Sigma$ by complex constants. Since $H^0(\Sigma, \Omega^2)$ is the cotangent space at Σ of the Teichmüller space $T_{p,k}$ of surfaces of the topological type of Σ (of genus p and with k boundary curves), and $H^0(\Sigma', \Omega^2)$ is the corresponding cotangent space for the surfaces of the type of Σ', i.e. of genus $2p + k - 1$ and without boundary, we can thus identify $T_{p,k}$ as a totally real submanifold of $T_{2p+k-1} = T_{2p+k-1,0}$. With these notations at hand, we can apply the notion of determinant described above to the operator built up from the Cauchy–Riemann operators in section I.4.

Let Σ be as before a Riemann surface with boundary and let g be a conformal metric on Σ. The operator P_g defined in (4.10) is an elliptic operator and we can define as above

$$\det P_g = (\det \ker P_g)^* \otimes (\det \ker P_g^*).$$

Since $\ker P_g = \ker \bar{\partial}_{-1}$ is independent of the metric g for a fixed conformal class (i.e. independent of the conformal parameter p, see (4.4)), we shall drop the symbol g and write $\det P_\Sigma$ for $\det P_g$.

The conformal metric g on Σ extends to a conformal metric g' on Σ' and we can define $P_{g'}$ and $\det P_{g'}$, which we shall denote $\det P_{\Sigma'}$.

In the following we shall also denote by $\ker \bar{\partial}_{n,\Sigma}, \ker \bar{\partial}^*_{n,\Sigma}$ (resp. $\ker \bar{\partial}_{n,\Sigma'}$, $\ker \bar{\partial}^*_{n,\Sigma'}$) the kernels of the Cauchy-Riemann operators relative to the metric g on Σ (resp. g' on Σ') since they only depend on the conformal class Σ of the metric g (resp. conformal class Σ' of the metric g') and not on the choice of the metric in the conformal class, as can be seen from (4.5) and (4.6).

With the choice of boundary conditions (4.20) and (4.22) for the operator P_g, we have

$$\det P_\Sigma = \det \bar{\partial}_{-1,\Sigma} = (\det \ker \bar{\partial}_{-1,\Sigma})^* \otimes (\det \ker \bar{\partial}^*_{-1,\Sigma}) \\ = \Lambda^{\max}(H^0(\Sigma, \Omega^2))^* \tag{7.52}$$

($\Lambda^{\max}$ denoting the maximal exterior power and

$$\det P'_\Sigma = \Lambda^{\max}(H^0(\Sigma', \Omega^2))^* = \det P_\Sigma \otimes_{I\!R} \mathbb{C} . \tag{7.53}$$

Likewise

$$\det \bar{\partial}_{0,\Sigma} = (\det \ker \bar{\partial}_{0,\Sigma})^* \otimes (\det \ker \bar{\partial}^*_{0,\Sigma}) \\ = i\, I\!R \otimes \Lambda^{\max}(H^0(\Sigma, \Omega))^* \tag{7.54}$$

and

$$\det \bar{\partial}_{0,\Sigma'} = \mathbb{C} \otimes \Lambda^{\max} H^0(\Sigma', \Omega) = \det \bar{\partial}_{0,\Sigma} \otimes_{I\!R} \mathbb{C}. \tag{7.55}$$

Let us also compare the eigenvalues of the $\bar{\partial}^*\bar{\partial}$ operators on Σ and Σ'. The important observation is that $\bar{\partial}^*\bar{\partial}$ is a real operator in the following sense: If ϕ is an n-differential on Σ' with

$$\phi(z) = \bar{\phi}(i(z)) \tag{7.56}$$

where $i : \Sigma' \to \Sigma'$ is the involution defined above, then also

$$\bar{\partial}^*_n \bar{\partial}_n \phi(z) = \overline{\bar{\partial}^*_n \bar{\partial}_n \phi}(i(z)). \tag{7.57}$$

Consequently, if we write for a general ϕ

$$\phi = (\phi^1 + \phi^2),$$

with

$$\phi^1(z) = \frac{1}{2}\left(\phi(z) + \bar{\phi}(i(z))\right)$$
$$\phi^2(z) = \frac{1}{2}\left(\phi(z) - \bar{\phi}(i(z))\right),$$

then

$$\bar{\partial}^*\bar{\partial}\phi = \bar{\partial}^*\bar{\partial}\phi^1 + \bar{\partial}^*\bar{\partial}\phi^2,$$

and if ϕ is an eigenvector,

$$\bar{\partial}^*\bar{\partial}\phi = \lambda\phi,$$

then ϕ^1 and ϕ^2 are eigenvectors with the same eigenvalue,

$$\bar{\partial}^*\bar{\partial}\phi^j = \lambda\phi^j, \quad j = 1,2.$$

Finally, if $\Gamma(\Sigma', K^n)$ denotes the smooth n-forms on Σ', and $\Gamma_{I\!R}(\Sigma', K^n)$ those which are invariant under i, i.e. satisfiy $\phi(z) = \bar{\phi}(i(z))$, then

$$\Gamma(\Sigma', K^n) = \Gamma_{I\!R}(\Sigma', K^n) \otimes_{I\!R} \mathbb{C}.$$

Thus, the eigenspaces for $\bar{\partial}^*\bar{\partial}$ operating on elements of $\Gamma(\Sigma', K^n)$ are the complexifications of those for $\bar{\partial}_n^*\bar{\partial}_n$ on $\Gamma_{I\!R}(\Sigma', K^n)$.

On the other hand, let $\Gamma(\Sigma, K^n)$ be the space of smooth n-forms

$$\psi = \psi^1 + i\psi^2$$

on Σ which are real on $\partial\Sigma$, i.e.

$$\psi^2 = 0 \text{ on } \partial\Sigma, \quad \text{locally given as } y = 0,$$

and satisfy also

$$\psi_y^1 = 0 \text{ on } \partial\Sigma.$$

If ψ is an eigenvector for $\bar{\partial}^*\bar{\partial}$ to the eigenvalue λ, i.e.

$$\bar{\partial}^*\bar{\partial}\psi = \lambda\psi$$

then one observes that ψ can be reflected to an n-form on Σ' which is invariant under i.

It follows from these observations that $\bar{\partial}_n^*\bar{\partial}_n$ has the same eigenvalues on Σ' and Σ, with the described boundary conditions on the latter, and that the eigenspaces on Σ' are the complexifications of those on Σ.

I.8 Chern classes of determinant bundles

Let us first recall some basic concepts from algebraic geometry; cf. [GH], [H]. We let M be a nonsingular quasiprojective variety. Later on, these concepts will be applied to the moduli space of Riemann surfaces of genus p. This space is not nonsingular, but since the singularities can be removed by passing to finite covers the theory will nevertheless be applicable. We let $\mathcal{O}$ be the sheaf of holomorphic functions on M, $\mathcal{O}^*$ of those which never vanish, $\mathcal{M}$ the sheaf of meromorphic functions, and $\mathcal{M}^*$ of the meromorphic functions which are not identically zero.

A divisor on M is a global section of the quotient sheaf $\mathcal{M}^*/\mathcal{O}^*$, i.e. an element of $H^0(M, \mathcal{M}^*/\mathcal{O}^*)$. Thus, if (U_α) is a cover of M by Zariski open subsets, then a <u>divisor</u> is given by a collection of not identically zero meromorphic functions f_α on U_α with

$$\frac{f_\alpha}{f_\beta} \in 0^*(U_\alpha \cap U_\beta). \tag{8.1}$$

Equivalently, this divisor is given as a formal linear combination of the zero and polar sets of the f_α

$$D = \sum_V ord_V(f_\alpha) \cdot V \tag{8.2}$$

which is well defined because of (8.1). Two divisors D_1 and D_2 defined by $(f_\alpha^1), (f_\alpha^2)$, respectively, are called linearly equivalent, if

$$\frac{f_\alpha^1}{f_\beta^1} = \frac{f_\alpha^2}{f_\beta^2} \quad \text{for all} \quad \alpha, \beta. \tag{8.3}$$

This is equivalent to saying that the collection $\frac{f_\alpha^1}{f_\alpha^2}$ defines a global meromorphic function on M.

We now turn to a notion under which linearly equivalent divisors are identified. A <u>holomorphic line bundle</u> L on M is defined by local trivializations

$$\phi_\alpha : L_{U_\alpha} \to U_\alpha \times \mathbb{C}.$$

The important objects are the transition functions

$$g_{\alpha\beta} := \frac{\phi_\alpha}{\phi_\beta} \in \mathcal{O}^*(U_\alpha \cap U_\beta) \tag{8.4}$$

which satisfy

$$\begin{aligned} g_{\alpha\beta}g_{\beta\alpha} &= 1 \\ g_{\alpha\beta}g_{\beta\gamma}g_{\gamma\alpha} &= 1. \end{aligned} \tag{8.5}$$

Conversely, a collection of transition functions satisfying (8.5) defines a <u>line bundle</u> L over M. If $(f_\alpha) \in \mathcal{O}^*(U_\alpha)$, then

$$\phi'_\alpha = f_\alpha \phi_\alpha$$

or

$$g'_{\alpha\beta} = \frac{f_\alpha}{f_\beta} g_{\alpha\beta} \qquad (8.6)$$

define the same line bundle, and conversely, two sets of transition functions of the same line bundle are related by such a collection $f_\alpha \in \mathcal{O}^*(U_\alpha)$. Thus, we can identify the set of line bundles on M with the cohomology group

$$H^1(M, \mathcal{O}^*) =: Pic M, \qquad (8.7)$$

the <u>Picard group</u> of M. We also recall

$$Div M := H^0(M, \mathcal{M}^*/\mathcal{O}^*). \qquad (8.8)$$

The constructions can be cohomologically interpreted as follows.

The short exact sequence

$$0 \to \mathcal{O}^* \to \mathcal{M}^* \to \mathcal{M}^*/\mathcal{O}^* \to 0 \qquad (8.9)$$

induces the long exact sequence

$$\begin{aligned} 0 \to & H^0(\mathcal{M}, \mathcal{O}^*) \to H^0(M, \mathcal{M}^*) \to H^0(M, \mathcal{M}^*/\mathcal{O}^*) \to \\ & H^1(M, \mathcal{O}^*) \to H^1(M, \mathcal{M}^*) \to \ldots \end{aligned} \qquad (8.10)$$

Now

$$H^1(M, \mathcal{M}^*) = 0 \qquad (8.11)$$

for nonsingular quasiprojective varieties. Therefore, the induced map

$$Div M \to Pic M \qquad (8.12)$$

is surjective, and linearly equivalent divisors are identified, i.e. those which only differ by a global meromorphic function, i.e. an element of $H^0(M, \mathcal{M}^*)$, this function being determined up to an element of $H^0(M, \mathcal{O}^*)$ which in our applications will be $\mathbb{C}^*$.

We now need to describe the concepts underlying the Grothendieck–Riemann–Roch theorem. The set of holomorphic vector bundles, on M, $Vect(M)$, is a semigroup under direct sum $\oplus$. In order to turn it into a group, we have to map

$$V - V_1 - V_2$$

I.8: Chern classes of determinant bundles 61

to 0, whenever
$$0 \to V_1 \to V \to V_2 \to 0$$
is exact. By making these identifications, we obtain a map

$$\text{Vect}(M) \to K(M)$$
$$V \to [V] \tag{8.13}$$

into a free abelian group $K(M)$, the Grothendieck group of M. We also need the Chow ring $A(M)$. A cycle of codimension r on M is an element of the free abelian group generated by the subvarieties of M. Thus, a cycle of codimension 1 is nothing but a divisor on M. If now N is a subvariety of M, then there exists a desingularization $\tilde{N}$ by a deep theorem of Hironaka, and a holomorphic map $f : \tilde{N} \to N$. We then call cycles D_1, D_2 rationally equivalent, if they are contained in some subvariety N and are of the form $D_1 = f_*C_1, D_2 = f_*C_2$, where $F : \tilde{N} \to N$ as before and C_1 and C_2 are linearly equivalent divisors in $\tilde{N}$. For $r = 1$, rationally equivalent cycles are the same as linearly equivalent divisors. The group of cycles of codimension r modulo rationally equivalent cycles is denoted by $A^r(M)$. The intersection

$$A^r(M) \times A^s(M) \to A^{r+s}(M),$$

the precise definition of which can be found in [H, pp. 426 f.], makes

$$A(M) := \oplus_{r=0}^{\dim M} A^r(M)$$

into a commutative graded ring with identity, the <u>Chow ring</u> of M.

We now return to holomorphic vector bundles over M. To each such bundle, we can associate its <u>Chern classes</u>. These are cohomology classes computed from the curvature matrix of a connection on the bundle, and since this construction can be carried out for an arbitrary complex, not necessary holomorphic bundle, they so far only reflect topological information about the bundle. If one introduces a Hermitian metric and uses the corresponding metric complex connection on a holomorphic vector bundle E, however, then its Chern classes $c_p(E)$ are of Hodge type (p,p), and in the present projective case, they are represented by the Poincaré duals of cycles (in the sense introduced for the definition of $A(M)$), and these cycles are determined up to rational equivalence. In particular, the first Chern class of a holomorphic line bundle L thus determines a divisor D up to linear equivalence, and the line bundle determined by D in turn is L. By abuse of notation (not distinguishing between a cohomology class and its Poincaré dual), we thus obtain maps

$$c_i : K(M) \to A^i(M), \quad 0 \leq i \leq \text{rank} E.$$

I.8: Chern classes of determinant bundles

We also need the Chern polynomial of $E \in K(M)$, with $m = rank E$,

$$c_t(E) := c_0(E) + \ldots + c_m(E)t^m$$
$$=: \prod_{i=1}^{m}(1 + a_i t),$$

where the a_i are formal symbols to be interpreted as 2-forms (note $c_0(E) = 1$). This in turn allows us to define the exponential Chern character

$$ch(E) := \sum_{i=1}^{m} e^{a_i}$$

with

$$e^a = 1 + a + \frac{1}{2}a^2 + \ldots.$$

One derives the expansion

$$ch(E) = m + c_1(E) + \frac{1}{2}(c_1^2(E) - 2c_2(E)) + \ldots$$

(formally putting $c_i(E) = 0$ for $i > m$). We also need the Todd class

$$td(E) = \prod_{i=1}^{m} \frac{a_i}{1 - e^{-a_i}},$$

with

$$\frac{a}{1 - e^{-a}} = 1 + \frac{1}{2}a + \frac{1}{12}a^2 + \ldots;$$

thus

$$td(E) = 1 + \frac{1}{2}c_1(E) + \frac{1}{12}(c_1^2(E) + c_2(E)) + \ldots.$$

We now look at functorial properties of these constructions. We let M, N be nonsingular quasiprojective varieties, and $f : M \to N$ be a proper holomorphic map which is everywhere of maximal rank $\dim N$. Thus, all fibers $f^{-1}(y), y \in N$, are regular. We then define an additive map

$$f_! : K(M) \to K(N)$$
$$E \to \sum_{i=0}^{rank E} (-1)^i R^i f_*(E).$$

In the case of interest to us below, the cohomology groups of the fibers of f occurring in the following equation have constant dimensions, and in this case

$$R^i f_*(E)_{|y} = H^i(f^{-1}(y), E_{|f^{-1}(y)}),$$

I.8: Chern classes of determinant bundles

the latter being the cohomology of the sheaf of sections of E over $f^{-1}(y)$. Finally, we let $T_{M/N}$ be the relative tangent bundle of f, i.e. the fiber over $x \in f^{-1}(y)$ is given by the tangent plane of $f^{-1}(y)$ at x. Under the assumption that f is a proper holomorphic map of nonsingular quasiprojective varieties everywhere of maximal rank – formulated above, but really needed only now – we then have the celebrated Riemann–Roch theorem of Grothendieck (generalizing the one of Hirzebruch) for $E \in K(M)$:

$$ch(f_!(E)) = f_*(ch(E) \cdot td(T_{M/N})) \quad \text{in} \quad A(N) \otimes Q. \qquad (8.14)$$

In other words, $f_!$ does not commute with ch, and the failure to commute is given by the Todd class. Thus, we have the following commutative diagram.

$$\begin{array}{c} K(M) \stackrel{ch \cdot td}{\longrightarrow} A(M) \otimes Q \\ f_! \downarrow \quad \quad \downarrow f_* \\ K(N) \stackrel{ch}{\longrightarrow} A(N) \otimes Q \end{array}$$

We now apply this to families of Riemann surfaces. Deviating slightly from the notation of sections I.3, I.4, we let Σ' be a compact Riemann surface without boundary of genus q, and we let T_q be the corresponding Teichmüller space. We define the mapping class group

$$\Gamma_q := \mathcal{D}(\Sigma')/\mathcal{D}_0(\Sigma'),$$

where $\mathcal{D}(\Sigma')$ denotes the orientation preserving diffeomorphisms of the oriented differentiable manifold underlying Σ', and $\mathcal{D}_0(\Sigma')$ those which are homotopic, and hence isotopic (see section I.4) to the identity. We then define the Riemann moduli space of surfaces of genus q as

$$M_q := T_q/\Gamma_q.$$

M_q is not a manifold but has quotient singularities at points corresponding to surfaces that admit nontrivial conformal automorphisms. A finite cover of M_q, however, is free of singularities, and the above results are therefore applicable to M_q. M_q is a quasiprojective variety by a result of Baily. Moreover, we let $\mathcal{M}_q$ be the universal modular curve, i.e. the fiber bundle over M_q with fiber over $y \in M_q$ given by the holomorphic curve determined by y. We then have a natural projection

$$\pi : \mathcal{M}_q \to M_q.$$

(Actually, there is again a technical problem here arising from the singularities of M_q. One either has to pass again to a finite cover of M_q, or else has to delete certain subvarieties from M_q and $\mathcal{M}_q$. As both spaces are noncompact anyway, this will not affect our subsequent reasoning).

We denote by $\tilde{\Omega}$ the relative cotangent bundle of π; thus the fibers of $\tilde{\Omega}$ are given by the cotangent bundles of the fibers of π. The relative tangent bundle of π in this notation is $\tilde{\Omega}^{-1}$. We now apply (8.14) to π, noting that the fibers are 1-dimensional, for a line bundle L on $\mathcal{M}_q$, and obtain

$$c_1(\pi_!(L)) = \pi_*(\frac{c_1(L)^2}{2} - \frac{c_1(L)c_1(\tilde{\Omega})}{2} + \frac{1}{12}c_1(\tilde{\Omega})^2). \qquad (8.15)$$

Here, π_* is given by integration over the fibers of π. Thus, the 4-form to which π_* is applied is converted into the 2-form on the left hand side of (8.15). From the above discussion, we recall that the left hand side of (8.15) is a divisor on M_q determined up to linear equivalence, and thus the holomorphic structure of the projected bundle $\pi_!(L)$ is completely determined by this divisor.

We want to apply the above arguments to the line bundles

$$\begin{aligned} L_n &:= \Lambda^{\max} H^0(\Sigma', \Omega^n), \ n > 1 \\ &= \Lambda^{\max} \pi_! \tilde{\Omega}^n = \Lambda^{\max} \pi_* \tilde{\Omega}^n, \end{aligned} \qquad (8.16)$$

remarking that $H^1(\Sigma', \Omega^n) = 0$ for $n > 1$, cf. (6.39), (6.30), and

$$\begin{aligned} L_1 &:= \Lambda^{\max} \pi_! \tilde{\Omega} = \Lambda^{\max} H^0(\Sigma', \Omega) \otimes \Lambda^{\max} H^1(\Sigma', \Omega) \\ &= \Lambda^{\max} H^0(\Sigma', \Omega) \otimes \Lambda^{\max} H^0(\Sigma', \mathbb{C})^* \quad \text{by (6.30)} \qquad (8.17) \\ &= \Lambda^{\max} H^0(\Sigma', \Omega) \otimes \mathbb{C}. \end{aligned}$$

We obtain

$$c_1(L_1) = \frac{1}{12} \pi_* c_1(\tilde{\Omega})^2 \qquad (8.18)$$

(L_1 is the so-called Hodge bundle, and

$$\lambda := c_1(L_1)$$

the Hodge divisor class), and more generally

$$c_1(L_n) = (6n(n-1) + 1)\lambda. \qquad (8.19)$$

In particular

$$c_1(L_n^* \otimes L_1^{6n(n-1)+1}) = 0, \qquad (8.20)$$

so that we obtain the holomorphic isomorphism

$$L_n \simeq L_1^{6n(n-1)+1}. \qquad (8.21)$$

We now recall

$$L_n^* = \det \bar{\partial}_{n-1} \qquad (8.22)$$

I.8: Chern classes of determinant bundles

and consequently we have on L_n^*, and hence also on L_n, the Quillen metric $\|\cdot\|_Q$, defined using the hyperbolic metrics on the fibers.

Now while (8.14) and (8.15) are equalities in the Chow ring, it turns out that if one uses the hyperbolic metrics on the fibers and the Quillen metric on L_n to compute the Chern forms, one obtains an equality for forms, due to Bismut and Freed [BF]:

$$c_1(L_n, \|\cdot\|_Q) = \int_{\mathcal{M}_q/M_q} ch(\tilde{\Omega}^n)\, td(\tilde{\Omega}^{-1}), \qquad (8.23)$$

where the integration is over the fibers as π (one integrates the 4-form part of the integrand over the fibers to obtain a 2-form on the base). Therefore, (8.21) becomes an isometry with respect to the Quillen metrics, hence

$$L_n = L_1^{6n(n-1)+1}, \qquad (8.24)$$

in particular

$$L_2 = L_1^{13}. \qquad (8.25)$$

As an interesting side remark, let us also mention that Wolpert [Wp] proved that if one uses the hyperbolic metrics on the fibers of the universal Teichmüller curve – as we are doing here – then

$$\int_{\mathcal{T}_q/T_q} c_1(\tilde{\Omega})^2 = \frac{1}{2\pi^2}\omega_{\text{WP}},$$

where ω_{WP} is the Kähler form of the Weil–Peterson metric. (Here, we are integrating as usual over the fibers of the universal Teichmüller curve $\mathcal{T}_q$ which is defined analogously to the universal modular curve.) Comparing with (8.23) and (8.18), we conclude

$$c_1(L_1, \|\cdot\|_Q) = \frac{1}{24\pi^2}\, \omega_{\text{WP}}$$

as an equality of forms. From (8.24), one then obtains

$$c_1(L_n, \|\cdot\|_Q) = \frac{6n(n-1)+1}{24\pi^2}\, \omega_{\text{WP}}.$$

We should point out, however, that one cannot interpret these equalities as equalities of cohomology classes on compactified moduli space; for this one needs a correction term determined by the Poincaré dual of the compactifying divisor.

I.9 Gaussian measures and random fields

In this section we recall some basic facts about infinite-dimensional measures and point out their relationships with quantum fields. We start with the finite-dimensional setting (e.g. [Bill]) and introduce the usual probabilistic terminology.

Let $(\Omega, \mathcal{F}, P)$ be a probability space. A <u>random vector</u> (or k-random vector) on Ω is a map $X : \Omega \to I\!R^k, \omega \to X(\omega) = (X_1(\omega), ..., X_k(\omega))$, which is $\mathcal{F}$-measurable. The <u>distribution</u> μ of X is the probability measure

$$\mu(A) := P((X_1, ..., X_k) \in A), \, A \in \mathcal{R}^k \qquad (9.1)$$

($\mathcal{R}^k$ the Borel σ-field over $I\!R^k$) and its Fourier transform

$$\varphi(t) := \int_{I\!R^k} e^{i(t,x)} \mu(dx), \, t \in I\!R^k, \qquad (9.2)$$

with $(t, x) := \sum_1^k t_\nu x_\nu$, is called the <u>characteristic function</u> of X and μ. It can be written in the form

$$\varphi(t) = E(e^{itX}),$$

where $E(Y)$ denotes integration of Y on Ω with respect to P and takes the name of <u>mean</u> or <u>expectation</u> of Y. Any characteristic function φ has the following three properties:
(1) φ is positive definite;
(2) $\varphi(0) = 1$;
(3) φ is uniformly continuous.

The above implication can be reversed, a relevant result due to Bochner (e.g. [Yos]):

Theorem 9.1 (Bochner): Let φ be any function on $I\!R^k$ which satisfies (1), (2) and (3). Then there exists a unique probability measure μ on $(I\!R^k, \mathcal{R}^k)$ such that

$$\varphi(t) = \int_{I\!R^k} e^{i(t,x)} \mu(dx).$$

We shall now consider Gaussian measures. A (centered) <u>Gaussian distribution</u> (or (centered) normal distribution) <u>on $(I\!R^k, \mathcal{R}^k)$</u> is any probability measure μ whose characteristic function is given by

$$\varphi(t) = e^{-\frac{1}{2}(t, Ct)} \qquad (9.3)$$

for some symmetric nonnegative definite matrix C, called the <u>covariance matrix</u>. μ can always be seen as the distribution of some random vector $X =$

I.9: Gaussian measures and random fields

$(X_1, \ldots, X_k)$ defined over a probability space. Random vectors which have a Gaussian distribution are called <u>Gaussian random vectors</u>.

Let us assume $C > 0$ and set $\Gamma := C^{-1}$. Then μ takes the explicit form

$$\mu(dx) = (2\pi)^{\frac{-k}{2}} (\det \Gamma)^{\frac{1}{2}} e^{-\frac{1}{2}(x, \Gamma x)} dx, \qquad (9.4)$$

dx being the Lebesgue measure on $\mathbb{R}^k$.

When $C = \Gamma = \mathbb{1}$ (identity matrix) μ is named <u>standard Gaussian distribution on $(\mathbb{R}^k, \mathcal{R}^k)$</u>.

Let us introduce the corresponding infinite-dimensional analog of these notions. $\mathcal{H}$ will indicate a real separable Hilbert space and $\mathcal{B}_\mathcal{H}$ the Borel field of $\mathcal{H}$. The scalar product on $\mathcal{H}$ will be denoted by $(\cdot, \cdot)$. By a <u>probability measure on $(\mathcal{H}, \mathcal{B}_\mathcal{H})$</u> we shall mean a positive Borel measure μ on $\mathcal{H}$ which is normalized, i.e. $\mu(\mathcal{H}) = 1$.

We shall consider only probability measures μ with respect to which the functional $f : \mathcal{H} \to \mathbb{R}$, $x \to f(x) := \|x\|$ ($\|\cdot\| = \sqrt{(\cdot, \cdot)}$), is square integrable. It is not hard to show that this requirement on μ is equivalent to the existence of an operator $C_\mu \in Cov(\mathcal{H})$, where

$Cov(\mathcal{H}) : \{C : \mathcal{H} \to \mathcal{H} \mid C \text{ linear, positive, self-adjoint trace class operator}\}$,

having the following properties:

$$(C_\mu x, y) = \int_\mathcal{H} (x, z)(z, y) \mu(dz) \quad \forall\, x, y \in \mathcal{H} \qquad (9.5)$$

and $tr C_\mu = \int_\mathcal{H} \|z\|^2 \mu(dz)$. Any element from the set $Cov(\mathcal{H})$ is called a <u>covariance operator</u> on $\mathcal{H}$.

We recall that a <u>positive definite functional</u> on $\mathcal{H}$ is map $\varphi : \mathcal{H} \to \mathbb{C}$ such that for any $x_1, \ldots, x_n \in \mathcal{H}$, $n = 1, 2, \ldots$, and any $c_1, \ldots, c_n \in \mathbb{C}$ one has:

$$\sum_{i,j=1}^n c_i \bar{c}_j \, \varphi(x_i - x_j) \geq 0. \qquad (9.6)$$

If μ is a probability measure on $\mathcal{H}$, its <u>characteristic functional</u> $\hat{\mu}$ is given by

$$\hat{\mu}(x) := \int_\mathcal{H} e^{i(x,y)} \mu(dy), x \in \mathcal{H} \qquad (9.7)$$

and $\hat{\mu}$ is easily seen to be positive definite and furthermore such that $\hat{\mu}(x) = 1$. The following result can be viewed as a generalization of Bochner's theorem for positive definite functionals on Hilbert spaces (see e.g. [Sko], [Kuo]).

I.9: Gaussian measures and random fields

Theorem 9.2 (Minlos–Prohorov–Sazonov): Let φ be any functional on $\mathcal{H}$ which satisfies the following three conditions:
(1) φ is positive definite;
(2) $\varphi(0) = 1$;
(3) $\forall \varepsilon > 0$, $\exists$ a covariance operator C_ε such that

$$1 - \operatorname{Re} \varphi(x) \le (C_\varepsilon x, x) + \varepsilon \quad \forall x \in \mathcal{H}. \tag{9.8}$$

Then φ is the characteristic functional of a probability measure on $\mathcal{H}$. Conversely any characteristic functional verifies (1), (2), and (3).

As an application let us consider the functional

$$\varphi(x) := e^{-\frac{1}{2}(x,Cx)} \quad \forall x \in \mathcal{H} \tag{9.9}$$

with $C \in Cov(\mathcal{H})$. φ satisfies trivially all the hypotheses of the previous theorem. For instance, since $1 - e^{-t} \le t \; \forall t \ge 0$, we have

$$1 - \operatorname{Re} \varphi(x) = 1 - e^{-\frac{1}{2}(x,Cx)} \le (x, \frac{C}{2}x) \quad \forall x \in \mathcal{H}$$

so condition (3) is verified. It follows that there is a probability measure μ on $(\mathcal{H}, \mathcal{B}_\mathcal{H})$ such that

$$\int_\mathcal{H} e^{i(x,y)} \mu(dx) = e^{-\frac{1}{2}(x,Cx)} \quad \forall x \in \mathcal{H}.$$

We now understand by a (centered) <u>Gaussian distribution on $(\mathcal{H}, \mathcal{B}_\mathcal{H})$</u> any probability measure μ whose characteristic functional has the form (9.9). At the same time it has become clear that the standard Gaussian distribution, well defined on $I\!\!R^k$, cannot be generalized directly to a probability measure on $(\mathcal{H}, \mathcal{B}_\mathcal{H})$, if $\dim \mathcal{H} = \infty$, since this would imply the identity operator to be trace class, a contradiction. There are however clever ways to handle the problem of defining the standard Gaussian distribution in the infinite-dimensional case.

Following L. Gross ([Gro]) we shall consider a locally convex real vector space L and its topological dual L^*. Let K be a finite-dimensional subspace of L^* and $\pi_K : L \to K^*$, $x \to \pi_K(x)$, the linear map defined by setting

$$\pi_k(x)y := <y, x>, \quad x \in L, \; y \in K \tag{9.10}$$

where $< \cdot , \cdot >$ denotes duality.

A <u>cylinder set</u> $\mathcal{C} \subset L$ is a set of the type $\mathcal{C} = \pi_K^{-1}(E)$ where E is a Borel set in K^* for a finite-dimensional subspace K of L^*. Let R denote the family of all cylinder sets and R_K the subfamily obtained fixing K; then R is a ring and R_K is a σ-ring. By definition a <u>cylinder set (normalized) measure</u> on L

I.9: Gaussian measures and random fields

is a set function $\mu : R \to [0, \infty)$ finitely additive over R and σ-additive over R_K for any subspace $K \subset L^*$ of finite dimension and such that moreover $\mu(L) = 1$.

As an example let us consider these definitions in the simplest case, when $L := \mathcal{H}$ is a real separable Hilbert space of arbitrary dimension. In this case $\mathcal{H}$ and $\mathcal{H}^*$ can be identified, the duality relation being now given by the Hilbertian scalar product $(\cdot, \cdot)$. A cylinder set in $\mathcal{H}$ has the form $\mathcal{C} = P^{-1}(E)$ where P is a finite-dimensional orthogonal projection and E is a Borel set in $P\mathcal{H}$. Let us set

$$\nu(\mathcal{C}) := (2\pi)^{-\frac{k}{2}} \int_E e^{-\frac{\|x\|^2}{2}} dx, \quad \mathcal{C} \in R, \tag{9.11}$$

where $k = \dim P\mathcal{H}$; then ν is a cylinder set measure called the (standard) <u>Gauss measure associated with $\mathcal{H}$</u>. However, it can be shown that when $\dim \mathcal{H} = \infty$ this cylinder set measure is not σ-additive on R, which is again a form of the troubles one encounters when trying to force some infinite-dimensional analog of the standard Gaussian distribution to live on $\mathcal{H}$.

To circumvent this unpleasant situation (the loss of σ-additivity implies e.g. loss of the possibility of performing usual operations such as interchanging between limits and integrals) one can adopt the following procedure. Pick up an operator $C \in Cov(\mathcal{H})$ and set

$$\|x\|_C := (x, Cx)^{\frac{1}{2}} \quad \forall x \in \mathcal{H}. \tag{9.12}$$

Under the assumption $\ker C = \{0\}$ this defines a new norm on $\mathcal{H}$. Let $\mathcal{H}_C$ denote the completion of $\mathcal{H}$ for the metric induced by $\|\cdot\|_C$. Let $\mathcal{P}$ be the collection of all the finite-dimensional projections of $\mathcal{H}$. It can be checked that $\|\cdot\|_C$ enjoys the following property:

$$\forall \varepsilon > 0 \quad \exists P_0 \in \mathcal{P} \text{ such that } \forall P \in \mathcal{P} \text{ orthogonal to } P_0.$$

One has

(m) $$\nu(\{x \mid \|Px\|_C > \varepsilon\}) < \varepsilon$$

where ν is the Gauss measure associated with $\mathcal{H}$. In general any norm on $\mathcal{H}$ which satisfies condition (m) is called a <u>measurable norm</u>.
From (9.12) is clear that the norm $\|\cdot\|_C$ is weaker than $\|\cdot\|$ (more important: not only $\|\cdot\|_C$ but any measurable norm has this property, as can be proven).

When $\mathcal{H}$ is infinite dimensional the Hilbertian norm $\|\cdot\|$ is not measurable, so $\|\cdot\|_C$ and $\|\cdot\|$ are never equivalent and therefore $\mathcal{H}_C$ contains $\mathcal{H}$ as a dense proper subset. The inclusion of $\mathcal{H}$ into $\mathcal{H}_C$ is realized by the Hilbert–Schmidt map $C^{\frac{1}{2}}$.

Let $\mathcal{H}_C^*$ be the topological dual of $\mathcal{H}_C$ (the space of linear continuous functionals on $\mathcal{H}_C$), and let $|_{\mathcal{H}}$:= restriction to $\mathcal{H}$. Then

$$|_{\mathcal{H}}: \mathcal{H}_C^* \longrightarrow \mathcal{H}^*$$
$$y \longrightarrow y|_{\mathcal{H}}$$

is a well-defined injective linear map by means of which we can consider $\mathcal{H}_C^*$ as a subset of $\mathcal{H}^* \simeq \mathcal{H}$, this last identification being given by the Riesz representation theorem. Moreover this subset is dense since $\mathcal{H}_C^*$ separates the points of $\mathcal{H}$. Therefore,

$$\mathcal{H}_C^* \hookrightarrow \mathcal{H} \hookrightarrow \mathcal{H}_C \qquad (9.13)$$

with inclusion maps having dense image. The inclusions (9.13) take the name of a <u>rigging of the Hilbert space $\mathcal{H}$</u> associated to $C \in Cov(\mathcal{H})$.

Remark: Since $\mathcal{H}_C$ is also a Hilbert space with respect to the scalar product $(x,y)_C = (C^{\frac{1}{2}}x, C^{\frac{1}{2}}y)$ it should be recalled that once the identification of $\mathcal{H}$ with $\mathcal{H}^*$ has been set up there is no room left for identifying $\mathcal{H}_C$ with $\mathcal{H}_C^*$ without falling into a contradiction. $\mathcal{H}_C^*$ must rather be seen as the subspace of vectors x of $\mathcal{H}$ such that $x \to (x, z)$ is continuous in the $\|\cdot\|_C$ topology. Let R_C denote the ring of cylinder sets of $\mathcal{H}_C$. Because of (9.13) this ring can be seen as a subfamily of cylinder sets of $\mathcal{H}$, and the set function ν_C obtained by restricting the Gauss measure ν to R_C defines a cylinder set measure on $\mathcal{H}_C$. The following theorem states that by this procedure cylinder sets behaving badly (with respect to ν) have been thrown away.

Theorem 9.3: ν_C is σ-additive on R_C. Furthermore ν_C can be uniquely extended to a σ-additive measure m_C on the σ-field generated by R_C which coincides with the Borel field $\mathcal{B}_C$ of $\mathcal{H}_C$.

We shall call the probability measure m_C on $(\mathcal{H}_C, \mathcal{B}_C)$ <u>the version on $\mathcal{H}_C$ of the Gauss measure associated with $\mathcal{H}$</u> (this latter being only a cylinder set measure).

Remark: To better appreciate the strength of property (m) let us recall a theorem of L. Gross (the result reported above being only a special case) which affirms the following: if $\bar{\mathcal{H}}$ is a completion of $\mathcal{H}$ w.r.t. any measurable norm, then there exists a probability measure on $(\bar{\mathcal{H}}, \bar{\mathcal{B}})$ ($\bar{\mathcal{B}}$ the Borel field of $\bar{\mathcal{H}}$) which is the version on the Banach space $\bar{\mathcal{H}}$ of the Gauss measure on $\mathcal{H}$. In this context the triple $(i, \mathcal{H}, \bar{\mathcal{H}})$, where $i : \mathcal{H} \hookrightarrow \bar{\mathcal{H}}$ denotes inclusion, is called an <u>abstract Wiener space</u>. Justifications for this name come from the following fact. Let $I := [0,1]$, $C := \{x \in C^0(I) \mid x(0) = 0\}$, $C' := \{x \in H^1(I) \mid x(0) = 0\}$. Completing the Hilbert space C' with respect to the sup norm $\|\cdot\|_\infty$ one gets the Banach space C. $\|\cdot\|_\infty$ is a measurable norm on C'

I.9: Gaussian measures and random fields

and the former construction gives on C, as probability measure, the classical Wiener measure μ_W. So classical Wiener space is associated to the triple (i, C', C). Moreover Wiener's famous result $\mu_W(C') = 0$ also generalizes to this framework in the sense that the probability measure of the Gross theorem supported by $\bar{\mathcal{H}}$ always assigns (in the infinite-dimensional case) measure zero to $\mathcal{H}$.

Let us now specify $\mathcal{H}$ as the Sobolev space $H^1 \equiv H^1(\mathbb{R}^d)$, $d \geq 1$. We recall that H^1 can be obtained by completing $\mathcal{S} \equiv \mathcal{S}(\mathbb{R}^d)$ (the Schwartz test function space) with respect to the Hilbert norm:

$$(x,x)_{H^1} := \int_{\mathbb{R}^d} \left((\nabla x)^2(u) + m^2 x^2(u)\right) du \qquad (9.14)$$

with $m > 0$. [†] We consider the Gauss measure ν associated with H^1 and let $\bar{\mathcal{H}}$ be a Hilbert space larger than H^1 (the choice of $\bar{\mathcal{H}}$ being immaterial in what follows) which supports a σ-additive realization of ν, here denoted by $\bar{\nu}$. As we know, this simply means that $i : H^1 \hookrightarrow \bar{\mathcal{H}}$ must be a (strictly) positive Hilbert–Schmidt inclusion with $i(H^1)$ dense in $\bar{\mathcal{H}}$. Let $\mathcal{S}' = \mathcal{S}'(\mathbb{R}^d)$ be the space of tempered distributions. Clearly $\bar{\mathcal{H}} \subset \mathcal{S}'$ so we can extend $\bar{\nu}$ to a probability measure μ_0 on $\mathcal{S}'$ by setting $\mu_0(A) := \bar{\nu}(A \cap \bar{\mathcal{H}}) \; \forall A \in \mathcal{B}(\mathcal{S}')$, the Borel σ-field of $\mathcal{S}'$.

In the mathematical physics literature on scalar quantum fields the probability measure μ_0 takes the name of <u>free Euclidean field measure (of mass m) on $\mathbb{R}^d$</u>. For every $x \in H^1$ the functional $y \to (x,y)_{H^1}$ on $\mathcal{S}'$ (where $(x,y)_{H^1} = \lim_{n\to\infty}(x,y_n)_{H^1}$, for some $(y_n) \subset \mathcal{S}$) is μ_0-measurable and the following holds:

$$\int_{\mathcal{S}'} e^{i(x,y)_{H^1}} \mu_0(dy) = e^{-\frac{1}{2}(x,x)_{H^1}} \quad \forall x \in H^1. \qquad (9.15)$$

We shall now sketch another equivalent way to describe the free Euclidean field measure μ_0 on $\mathcal{S}'$. The idea is to build up infinite-dimensional measures (in particular of Gaussian type) directly on the dual of a nuclear space. Denoting as previously by L a real vector space we recall some basic definitions ([GV], [Hida]). Assume there is given on L a countable family of inner products $\{(\cdot,\cdot)_n\}_{n \geq 0}$ which are compatible, i.e. if $(u_k) \subset L$ is such that $\|u_k\|_m \to 0$ ($\|\cdot\|_m := \sqrt{(\cdot,\cdot)_m}$) and is Cauchy in $\|\cdot\|_n$ then $\|u_k\|_n \to 0$. L is called a <u>countable Hilbert space</u> if it is complete in the metric topology defined by the norm $\|u\| := \sum_{n \geq 0} 2^{-n}(\|u\|_n/(1 + \|u\|_n))$ (in particular L will be locally convex). Set $L_n :=$ completion of L in the norm $\|\cdot\|_n$; then completeness of L w.r.t. the previous metric is equivalent to $L = \bigcap_{n \geq 0} L_n$.

[†] Sobolev spaces were already introduced in I.3. For notational convenience their elements will be denoted in the present section by small letters (e.g. we reserve the capital letter X for embeddings).

Notice that it is always possible to choose norms such that $\|\cdot\|_0 \leq \|\cdot\|_1 \leq \ldots$ without changing the topology, so we can assume $L_0 \supset L_1 \supset \ldots$.

A <u>nuclear space</u> is a countable Hilbert space which has the further property: $\forall m \; \exists n > m$ such that the injection map $i_m^n : L_n \to L_m$ is trace class.

Previously we have considered cylinder set measures on L, while now we shall consider cylinder set measures on L^*, the topological dual, where we are thinking of L as a countable Hilbert space. Indeed, the definition of a cylinder set in L^* can be obtained directly from (9.10) exchanging the roles of L and L^* and taking K as a finite-dimensional subspace of L. The definition of cylinder set measure on L^* then follows in the same way and we can ask the question under which conditions this measure is σ-additive. A first answer is that nuclearity of the linear space L is a necessary condition ([GV], chap. IV). Let μ be a probability measure on L^*. Then its characteristic functional $\hat{\mu}$ is defined as in (9.7), but now the Hilbertian scalar product is replaced by the duality relation, therefore

$$\hat{\mu}(x) = \int_{L^*} e^{i<x,y>} \mu(dy), x \in L. \qquad (9.16)$$

The notion of positive definite functional on L being as in (9.6) (where now $x_1, \ldots, x_n \in L$) we can state the generalization of Bochner's theorem to this setting:

Theorem 9.4 (Minlos): Let L be a nuclear space and φ a functional on L. Then the following three conditions:
(1) φ is positive definite;
(2) $\varphi(0) = 1$;
(3) φ is continuous w.r.t. the metric topology of L, are necessary and sufficient for the existence of a unique probability measure on L^* which has φ as characteristic functional.

Let $B = B(x, x'), x, x' \in L$, L a nuclear space, be a continuous (strictly) positive bilinear form and set

$$\varphi(x) := e^{-\frac{1}{2}B(x,x)} \qquad \forall x \in L. \qquad (9.17)$$

Since φ verifies the conditions of Minlos's theorem there is a probability measure μ on $(\mathcal{B}_{L^*}, L^*)$ such that $\hat{\mu}(x) = \varphi(x)$. Any measure of this type is called <u>a (centered) Gaussian distribution</u> on $(L^*, \mathcal{B}_{L^*})$ and B is called the <u>covariance functional</u> of the <u>Gaussian distribution</u>.

Within this framework we recover the free Euclidean field measure of mass m on $\mathbb{R}^d$ by choosing as nuclear space $L := \mathcal{S}$ and as form $B(x, x') := ((-\Delta + m^2)^{-1}x, x')_{L^2(\mathbb{R}^d)} = (x, x')_{H^{-1}(\mathbb{R}^d)}$. Indeed such a form satisfies all

I.9: Gaussian measures and random fields

the above requirements and therefore there is a probability measure $\tilde{\mu}_0$ on $\mathcal{S}'$ such that

$$\int_{\mathcal{S}'} e^{i<x,y>} \tilde{\mu}_0(dy) = e^{-\frac{1}{2}(x,x)_{H^{-1}}} = e^{-\frac{1}{2}(\psi,\psi)_{H^1}} \qquad (9.18)$$

where $\psi \equiv (-\Delta+m^2)^{-1}x$ so that $(x,x)_{H^{-1}} = (\psi,(-\Delta+m^2)\psi)_{L^2} = (\psi,\psi)_{H^1}$; moreover $(\psi,y)_{H^1} = \left((-\Delta+m^2)\psi,y\right)_{L^2} = (x,y)_{L^2} = <x,y>$ for $y \in L^2$ while for arbitrary y the equality holds by approximation since L^2 is dense in H^{-1}. By uniqueness, comparing (9.18) with (9.15), it follows that $\tilde{\mu}_0$ is exactly the free Euclidean field measure, i.e. $\tilde{\mu}_0 = \mu_0$.

Let us remark that in the above presentation we have chosen to put emphasis on measures rather than on processes, hence, before ending this section, we shall fill up the gap by giving a brief description of the connections between these concepts. We shall start by recalling the concept of stochastic process, as a tool to introduce later on random fields. The notion of stochastic or random process generalizes that of k-random vector given at the beginning of this section. Namely a (real-valued) stochastic process (or random process) is a family $(X_t)_{t \in T}$ of measurable (real-valued) functions over some probability space $(\Omega, \mathcal{F}, P)$, where the index set T can be discrete or continuous.

In the following we shall assume $T \subset \mathbb{R}$, i.e. we shall look at the index set as a subset of the real line. Given a random process $(X_t)_{t \in T}$ one can define a set of functions $\{F_{t_1,\ldots,t_n}(x_1,\ldots,x_n)\}$, where $x_i \in \mathbb{R}$, $t_i \in \mathbb{R}$ and $t_1 < t_2 < \ldots < t_n$, by

$$F_{t_1,\ldots,t_n}(x_1,\ldots,x_n) := P(X_{t_1} \leq x_1, \ldots, X_{t_n} \leq x_n) \qquad (9.19)$$

called the finite-dimensional distributions of $(X_t)_{t \in T}$. They verify

$$\lim_{x_k \to \infty} F_{t_1,\ldots,t_n}(x_1,\ldots,x_n) = F_{t_1,\ldots,\hat{t}_k,\ldots,t_n}(x_1,\ldots,\hat{x}_k,\ldots,x_n) \qquad (9.20)$$

($\hat{}$ denotes omission) which is known as the consistency property. At this point there is a famous classical theorem of Kolmogoroff (see any book on random processes) which states that starting from any, a priori given, family of functions $\{F_{t_1,\ldots,t_n}(x_1,\ldots,x_n)\}$ satisfying the consistency property and some other simple properties (whose necessity can be seen from (9.19)) it is possible to manufacture a probability space and a random process which has such a family as its finite-dimensional distributions.

Generally speaking Kolmogoroff's solution of the existence problem consists first in choosing $\Omega \equiv \mathbb{R}^T$ and $\mathcal{F} \equiv \mathcal{B}(\mathbb{R}^T)$ (the Borel σ-algebra generated by cylinder sets), hence in considering the family of measures $(P_{t_1,\ldots,t_n})$ on cylinder sets determined by

$$P_{t_1,\ldots,t_n}(\omega_{t_1} \leq x_1,\ldots,\omega_{t_n} \leq x_n) \equiv F_{t_1,\ldots,t_n}(x_1,\ldots,x_n)$$

and defining
$$P((\omega_{t_1},...,\omega_{t_n}) \in B) \equiv P_{t_1,...,t_n}(B)$$
on cylinder sets. Then "consistency" is used to prove that P extends to a probability measure on $(\mathbb{R}^T, \mathcal{B}(\mathbb{R}^T))$ while the process $(X_t)_{t \in T}$ is realized as the so-called coordinate process, i.e. $X_t(\omega) \equiv \omega_t$, $\omega \in \mathbb{R}^T$.

Let us notice that the chosen sample space is a huge one so it is natural to address the question in which cases it is possible to accomodate the support of P on some more suitable measurable subset of $\mathbb{R}^T$. This depends on the problem at hand; one possibility is furnished by Kolmogoroff's criterion which gives simple conditions (involving only 2-dimensional distributions!) which are sufficient to ensure that P is concentrated on $C(T, \mathbb{R}) \subset \mathbb{R}^T$ (see e.g. [Bill]).

We shall now introduce generalized processes. Let L be a locally convex topological space and $(\Omega, \mathcal{F}, P)$ a probability space. A generalized stochastic process (or a generalized random field, when L is realized by a space of functions in more than one variable) is a map $\varphi : L \to \{$random variables on $(\Omega, \mathcal{F}, P)\}$ satisfying the following two conditions:

(1) $\varphi(\alpha x + \beta x') = \alpha \varphi(x) + \beta \varphi(x')$ $\forall x, x' \in L$ and $\alpha, \beta \in \mathbb{R}$, with probability 1.

(2) $\varphi(x)$ is continuous in the following sense: if $x_{k_i} \to x_k$ in L as $i \to \infty$ for $k = 1, ..., n$ then the vector $(\varphi(x_{1_i}), ..., \varphi(x_{n_i})) \to (\varphi(x_1), ..., \varphi(x_n))$ in law as $i \to \infty$.

Notice that classical stochastic processes can be viewed as particular kinds of generalized stochastic processes.

To stress the connections with the previous part of this section let us recall the notion of Gaussian generalized stochastic process (shortly G.g.s.p.). A G.g.s.p. is a generalized process such that for arbitrary linearly independent elements $x_1, ..., x_n \in L$, $n = 1, 2, ...$, the random vector $(\varphi(x_1), ..., \varphi(x_n))$ is Gaussian. A G.g.s.p. is said to be centered (or of zero mean) if $E(\varphi(x)) = 0$ $\forall x \in L$. We shall consider only centered G.g.s.p.s. Given a G.g.s.p. $\varphi \equiv (\varphi(x)|x \in L)$ we can build up a covariance functional $B : L \times L \to \mathbb{R}$ by setting $B(x, x') \equiv E(\varphi(x)\varphi(x'))$; then by (9.17) and Minlos' theorem we obtain an associated Gaussian probability measure on L^*. Conversely for any covariance functional B, i.e. for any continuous strictly positive bilinear form on L, there is a G.s.p. φ such that $E(\varphi(x)\varphi(x')) = B(x, x')$.

If μ_0 is the free Euclidean field measure discussed before, then the corresponding G.g.s.p. φ is called the free Euclidean field (of mass m, over $\mathbb{R}^k$). It takes values, by the above realization of L, in the (real) distribution space $S'(\mathbb{R}^k)$.

I.10 Functional quantization of the Høegh–Krohn and Liouville models on a compact surface.

In this section S will denote a connected compact smooth surface without boundary having arbitrary genus. For g a fixed Riemannian metric on S, $m_g(dx)$ will be the associated Riemannian volume form. Let us consider on (S, g) the free field measure of unit mass, i.e. the Gaussian measure μ_0 on $\mathcal{D}' \equiv \mathcal{D}'(S)$, the space of distributions on S, for which

$$\int_{\mathcal{D}'} e^{i<\varphi,u>} \mu_0(d\varphi) = e^{-\frac{1}{2}(u,(1-\Delta_g)^{-1}u)} \tag{10.1}$$

with $u \in \mathcal{D} \equiv \mathcal{D}(S)$, Δ_g the Laplace–Beltrami operator on (S, g), $<\varphi, u> = \varphi(u)$ the duality relation between $\mathcal{D}'$ and $\mathcal{D}$, given by $(\cdot, \cdot)$, the $L^2(S, m_g)$-scalar product. We remark that μ_0 depends on g. Choose an arbitrary $\psi \in \mathcal{D}'$ and consider the new measure $\mu_0^\psi(\cdot) \equiv \mu_0(\cdot - \psi)$ on $\mathcal{D}'$. Then the Cameron–Martin–Girsanov–Maruyama theorem (see e.g. [Kuo]) tells us that μ_0^ψ is absolutely continuous with respect to μ_0 iff $(1 - \Delta_g)^{\frac{1}{2}}\psi \in L^2(S, m_g)$ (i.e. $\psi \in H^1(S)$). Moreover if this condition is satisfied the Radon–Nikodým derivative is given by

$$\frac{d\mu_0^\psi}{d\mu_0}(\varphi) = e^{<\varphi,(1-\Delta_g)\psi>} e^{-\frac{1}{2}(<\psi,(1-\Delta_g)\psi>)}. \tag{10.2}$$

Now let $p(t, x, y)$ be the heat kernel on (S, g) (dependence on g is dropped from the notation) and set

$$G(x, y) \equiv \int_0^\infty e^{-t} p(t, x, y) dt \tag{10.3}$$

so that $G(x, y) = (1 - \Delta_g)^{-1}(x, y)$, i.e. $G(x, y)$ is the Green function for the equation $(1 - \Delta_g)f = h$ on (S, g). If we choose $\psi \equiv G_x \in \mathcal{D}'$ (where $G_x(u) \equiv \int_S u(y) G(x, y) m_g(dy)$), then the measures $\mu_{0,x}(\cdot) \equiv \mu_0^{G_x}(\cdot)$ and μ_0 are singular for each $x \in S$ since $(1 - \Delta_g)^{\frac{1}{2}} G_x \notin L^2(S, m_g)$. For $\alpha \in \mathbb{R}$ we set $\mu_{0,x}^\alpha(\cdot) \equiv \mu_0^{\alpha G_x}(\cdot)$ and consider the new measure

$$\mu_0^\alpha(\cdot) \equiv \int_S \mu_{0,x}^\alpha(\cdot) m_g(dx). \tag{10.4}$$

We shall prove the following

Theorem 10.1: For $\alpha^2 < 4\pi$, μ_0^α is absolutely continuous with respect to μ_0, with $\frac{d\mu_0^\alpha}{d\mu_0} \geq 0$ and in $L^2(\mu_0)$.

I.10: Functional quantization

Remark: This result is not the best possible. Indeed recently Kusuoka [Ku] (see also [Kah]) has shown that absolute continuity holds for the larger interval $\alpha^2 < 8\pi$ (see also the Remark after Prop. below).

To prove Theorem 10.1 we shall proceed in several steps. Before doing this let us recall that the statement of Theorem 10.1 was originally proven in [AHK1] for the case of a 2-dimensional globally flat geometry in relation to the construction of a model of non-Gaussian scalar field theory known as Høegh-Krohn model or exponential interaction model (see [HK], [AHK1]). We shall come back to this later on. Let us first introduce for $\varepsilon > 0$

$$G_\varepsilon(x,y) \equiv \int_\varepsilon^{+\infty} e^{-t} p(t,x,y) dt \tag{10.5}$$

and consider the family of measures $(\mu_{0,x,\varepsilon}^\alpha)_{\varepsilon > 0}$, where $\mu_{0,x,\varepsilon}^\alpha(\cdot) \equiv \mu_0^{\alpha G_{\varepsilon,x}}(\cdot)$. Since G_ε is smooth we have $(1-\Delta_g)^{-\frac{1}{2}} G_{\varepsilon,x} \in L^2(S, m_g)$. Therefore the measures $\mu_{0,x,\varepsilon}^\alpha$ and also $\mu_{0,,\varepsilon}^\alpha \equiv \int_S \mu_{0,x,\varepsilon}^\alpha m_g(dx)$ are absolutely continuous with respect to μ_0, and we have from (10.2)

$$\begin{aligned}\frac{d\mu_{0,\varepsilon}^\alpha}{d\mu_0}(\varphi) &= \int_S \frac{d\mu_{0,x,\varepsilon}^\alpha}{d\mu_0}(\varphi) m_g(dx) \\ &= \int_S e^{\alpha \langle \varphi, (1-\Delta_g)G_{\varepsilon,x}\rangle} e^{-\frac{\alpha^2}{2}(\langle G_{\varepsilon,x},(1-\Delta_g)G_{\varepsilon,x}\rangle)} m_g(dx).\end{aligned} \tag{10.6}$$

We notice that $((1-\Delta_g)G_{x,\varepsilon})(y) = e^{-\varepsilon} p(\varepsilon,x,y)$, since $p(t,x,y)$ satisfies the equation $\frac{\partial p}{\partial t} = \Delta_g p$ on (S,g). Let us introduce the following notation:

$$\chi_\varepsilon^x(y) \equiv e^\varepsilon p(\varepsilon,x,y)$$
$$\varphi_\varepsilon(x) \equiv \varphi(\chi_\varepsilon^x) \quad x \in S, \; \varphi \in \mathcal{D}'.$$

We note that $\langle \chi_\varepsilon^x, (1-\Delta_g)^{-1}\chi_\varepsilon^y \rangle = \langle G_{\varepsilon,x}, (1-\Delta_g)G_{\varepsilon,y}\rangle$ for all $x, y \in S$. We set

$$: e^{\alpha \varphi_\varepsilon(x)} : \equiv e^{\alpha \varphi_\varepsilon(x)} e^{-\frac{\alpha^2}{2} E_{\mu_0}(\varphi_\varepsilon(x)^2)}$$

so that $E\left(: e^{\alpha \varphi_\varepsilon(x)} :\right) = 1$. We note that

$$: e^{\alpha \varphi_\varepsilon(x)} : \equiv e^{\alpha \varphi_\varepsilon(x)} e^{-\frac{\alpha^2}{2} \langle G_{\varepsilon,x}, (1-\Delta_g)G_{\varepsilon,x}\rangle}.$$

From the definitions above we have then

$$\begin{aligned}: e^{\alpha \varphi_\varepsilon(x)} : &= e^{\alpha \langle \varphi, \chi_\varepsilon^x\rangle} e^{-\frac{\alpha^2}{2}\langle G_{\varepsilon,x}(1-\Delta_g)G_{\varepsilon,x}\rangle} \\ &= e^{\alpha \langle \varphi, (1-\Delta_g)G_{\varepsilon,x}\rangle} e^{-\frac{\alpha^2}{2}\langle G_{\varepsilon,x}(1-\Delta_g)G_{\varepsilon,x}\rangle}.\end{aligned} \tag{10.7}$$

Hence from (10.6)

$$\frac{d\mu_{0,\varepsilon}^\alpha}{d\mu_0}(\varphi) = U_{\varepsilon,\alpha}(\varphi) \tag{10.8}$$

I.10: Functional quantization

with
$$U_{\varepsilon,\alpha}(\varphi) = \int_S :e^{\alpha\varphi_\varepsilon(x)}: m_g(dx). \tag{10.9}$$

We remark that $U_{\varepsilon,\alpha} \geq 0$ μ a.e. From (10.1) and (10.7) we get

$$E_{\mu_0}\left(:e^{\alpha\varphi_\varepsilon(x)}::e^{\alpha\varphi_\varepsilon(y)}:\right) = e^{\alpha^2 <G_{\varepsilon,x}(1-\Delta_g)G_{\varepsilon,y}>} \tag{10.10}$$

for all $\varepsilon > 0$, $x, y \in S$. Let

$$:\varphi_\varepsilon(x)^n: \equiv \sum_{k=0}^{[n/2]} \frac{n!}{k!(n-2k)!}\varphi_\varepsilon(x)^{n-2k}(-\frac{1}{2}<G_{\varepsilon,x},(1-\Delta_g)G_{\varepsilon,x}>)^k. \tag{10.11}$$

One sees easily (using properties of Gaussian measures) that

$$E_{\mu_0}(:\varphi_\varepsilon(x)^n::\varphi_{\varepsilon'}(y)^m:) = \delta_{nm}n!E_{\mu_0}(\varphi_\varepsilon(x)\varphi_{\varepsilon'}(y))^n. \tag{10.12}$$

In particular $:\varphi_{\varepsilon'}(x)^n:$ is orthogonal to $:\varphi_\varepsilon(y)^m:$ for all $n \neq m$. Using this we see that

$$:e^{\alpha\varphi_\varepsilon(x)}: = L^2_{(\mu_0)} \lim_{N\to\infty} \sum_{n=0}^N \frac{\alpha^n}{n!} :\varphi_\varepsilon(x)^n:. \tag{10.13}$$

Set
$$U_{\varepsilon,\alpha,n}(\varphi) \equiv \frac{\alpha^n}{n!}\int_S :\varphi_\varepsilon(x)^n: m_g(dx). \tag{10.14}$$

Lemma 10.1: For any $\varepsilon, \varepsilon' > 0$ we have

$$E(\varphi_{(\varepsilon\vee\varepsilon')}(x)^2) \leq E(\varphi_\varepsilon(x)\varphi_{\varepsilon'}(x)) \leq E(\varphi_{(\varepsilon\wedge\varepsilon')}(x)^2). \tag{10.15}$$

Proof: We have

$$E(\varphi_\varepsilon(x)\varphi_{\varepsilon'}(x)) = \int_{S\times S} \chi^x_\varepsilon(y)G(y,y')\chi^x_{\varepsilon'}(y')dydy'$$

$$= e^{-(\varepsilon+\varepsilon')}\int_{S\times S} p(\varepsilon,x,y)G(y,y')p(\varepsilon',x,y')dydy'.$$

But $p(t,x,y) \geq 0$ is a decreasing function in t. Thus

$$E(\varphi_{(\varepsilon\vee\varepsilon')}(x)^2) = e^{-(2(\varepsilon\vee\varepsilon')}\int_{S\times S} p(\varepsilon\vee\varepsilon',x,y)G(y,y')p(\varepsilon\vee\varepsilon',x,y')dydy'$$

$$\leq e^{-(\varepsilon+\varepsilon')}\int_{S\times S} p(\varepsilon,x,y)G(y,y')p(\varepsilon',x,y')dydy'$$

$$= E(\varphi_\varepsilon(x)\varphi_{\varepsilon'}(x)),$$

which proves the first inequality. The other one is proven in a similar way.
∎

Lemma 10.2: For $\varepsilon' > \varepsilon > 0$:

$$E(:\varphi_\varepsilon(x)^n::\varphi_{\varepsilon'}(x)^n) \geq E\left((:\varphi_{\varepsilon'}(x)^n:)^2\right).$$

Proof: This is immediate from (10.12) and Lemma 10.1. ∎

Lemma 10.3: For $\varepsilon' > \varepsilon > 0$

$$E_{\mu_0}\left((:\varphi_\varepsilon(x)^n:-:\varphi_{\varepsilon'}(x)^n:)^2\right) \leq n!\left[\left(E_{\mu_0}(\varphi_\varepsilon(x)^2)\right)^n - \left(E_{\mu_0}(\varphi_{\varepsilon'}(x)^2)\right)^n\right].$$

Proof: Set $a \equiv :\varphi_\varepsilon(x)^n:$, $b \equiv :\varphi_{\varepsilon'}(x)^n:$. Then $E_{\mu_0}(a-b)^2 = E_{\mu_0}(a^2) - 2E_{\mu_0}(ab) + E_{\mu_0}(b^2) \leq E_{\mu_0}(a^2) - E_{\mu_0}(b^2)$, since $E_{\mu_0}(ab) \geq E_{\mu_0}(b^2)$, by Lemma 10.2. From this the result follows, using (10.12). ∎

Using this Lemma we can easily prove the following:

Prop. 10.1: For any $\alpha \in \mathbb{R}$, $n \in \mathbb{N}$ $(U_{\varepsilon,\alpha,u})\varepsilon > 0$ converges in $L^2(\mu_0)$ as $\varepsilon \downarrow 0$.

Proof: Let $\varepsilon' > \varepsilon > 0$, then by Lemma 10.3 we have

$$\|U_{\varepsilon,\alpha,n} - U_{\varepsilon',\alpha,n}\|^2 \leq \frac{\alpha^{2n}}{n!}[H_n(\varepsilon) - H_n(\varepsilon')], \qquad (10.16)$$

where $H_n(\varepsilon) \equiv \int_S E_{\mu_0}(\varphi_\varepsilon(x)^2))^n m_g(dx)$. $H_n(\varepsilon)$ is continuous on $(0, \varepsilon_0]$ for any $\varepsilon_0 > 0$. Moreover the lim of $H_n(\varepsilon)$ for $\varepsilon \downarrow 0$ exists, since by Lemma 10.3 $H_n(\varepsilon)$ is increasing as ε decreases and is bounded by

$$\int_S G(x,x)^n m_g(dx) < \infty, \qquad (10.17)$$

since $A(x,y) \equiv G(x,y) + \frac{1}{2\pi}\ln d(x,y)$ is smooth on $S \times S$ (as follows from standard Jacobi field estimates, see e.g. [J4]). Hence we have

$$G(x,y) \leq C - \frac{1}{2\pi}\ln d(x,y) \qquad (10.18)$$

for some $C > 0$, which proves (10.17), the logarithmic singularity for $x = y$ being integrable. $H_n(\cdot)$ is thus a continuous function on $[0, \varepsilon_0]$, for $\varepsilon_0 > \varepsilon', \varepsilon$, and from (10.16) we deduce then that $U_{\varepsilon,\alpha,n}$ converges in $L^2(\mu_0)$ as $\varepsilon \downarrow 0$. ∎

I.10: Functional quantization

Let us call $U_{\alpha,n}$ the element of $L^2(\mu_0)$ such that, in the $L^2(\mu_0)$-sense,

$$U_{\alpha,n} = \lim_{\varepsilon \downarrow 0} U_{\varepsilon,\alpha,n}. \tag{10.19}$$

We then have

$$\begin{aligned}\|U_{\alpha,n}\|^2 &= \frac{(\alpha)^{2n}}{(n!)^2} E_{\mu_0}\left(|\int dm_g(x):\varphi(x)^n:|^2\right) \\ &= \frac{(\alpha)^{2n}}{(n!)} \int G(x,y) m_g(dx) m_g(dy).\end{aligned} \tag{10.20}$$

Lemma 10.4:
(i) For any $\varepsilon > 0$, $x,y \in S$: $<G_{\varepsilon,x},(1-\Delta_g)G_{\varepsilon,y}> \leq G(x,y)$
(ii) For $\alpha^2 < 4\pi$

$$\int_{S\times S} e^{\alpha^2 G(x,y)} m_g(dx)\, m_g(dy) < \infty$$

Proof: (i) is a direct consequence of the inequality

$$\int_{S\times S} p(\varepsilon,x,z)p(t,z,z')p(\varepsilon,z',y)m_g(dz)m_g(dz') \leq p(t,x,y) \tag{10.21}$$

which holds for any $t > 0$, as seen e.g. by expansion of the heat kernel in eigenfunctions of $-\Delta_g$.
(ii) From (10.18) we have:

$$\int_{S\times S} e^{\alpha^2 G(x,y)} m_g(dx)m_g(dy) \leq e^C \int_{S\times S} (d(x,y))^{-\beta} m_g(dx)m_g(dy), \tag{10.22}$$

with $\beta \equiv \frac{\alpha^2}{2\pi}$.

Since by assumption $\beta < 2 = \dim S$ and S is compact, for every $y \in S$

$$\int_S (d(x,y))^{-\beta} m_g(dx) < \infty$$

and the conclusion follows from (10.22). ∎

Lemma 10.5: For any $\varepsilon > 0$

$$\sum_n \|U_{\varepsilon,\alpha,n}\|^2 = \int_{S\times S} e^{\alpha^2 <G_{\varepsilon,x},(1-\Delta_g)G_{\varepsilon,x}>} m_g(dx)m_g(dy).$$

For $\alpha^2 < 4\pi$ (with $U_{0,\alpha,n} \equiv U_{\alpha,n}$) and any $\varepsilon \geq 0$:

$$\sum_n \|U_{\varepsilon,\alpha,n}\|^2 \leq \sum_n \|U_{\alpha,n}\|^2 = \int_{S\times S} e^{\alpha^2 G_{x,y}} m_g(dx)m_g(dy).$$

Proof: We have for $\varepsilon > 0$:

$$\begin{aligned}
\|U_{\varepsilon,\alpha,n}\|^2 &= E_{\mu_0}\left[|\int m_g(dx)\frac{\alpha^n}{n!} :\varphi_\varepsilon(x)^n :|^2\right] \\
&= \frac{(\alpha)^{2n}}{(n!)^2}\left[|\int m_g(dx) :\varphi_\varepsilon(x)^n :|^2\right] \\
&= \frac{(\alpha)^{2n}}{(n!)^2} \int m_g(dx)m_g(dy) E_{\mu_0}(:\varphi_\varepsilon(x)^n::\varphi_\varepsilon(y)^n:) \\
&= \frac{(\alpha)^{2n}}{n!} \int E_{\mu_0}([\varphi_\varepsilon(x)\varphi_\varepsilon(y)]^n) m_g(dx)m_g(dy),
\end{aligned} \quad (10.23)$$

where we have used (10.12). Hence by (10.23) and Lemma 10.4,

$$\sum_n \|U_{\varepsilon,\alpha,n}\|^2 = \int_{S\times S} e^{\alpha^2 <G_{\varepsilon,x},(1-\Delta_g)G_{\varepsilon,x}>} m_g(dx)m_g(dy)$$

$$\leq \int_{S\times S} e^{\alpha^2 G(x,y)} m_g(dx)m_g(dy).$$

The r.h.s. is finite, for $\alpha^2 < 4\pi$, by Lemma 10.4 (ii). The rest follows from the definition of $U_{\alpha,n}$, (10.19) and (10.20). ■

We have for any $\varepsilon > 0$ and also for $\varepsilon = 0$, $\alpha^2 < 4\pi$ that $\sum_n U_{\varepsilon,\alpha,n}$ converges in $L^2(\mu_0)$. In fact this follows from

$$\left\|\sum_{n=0}^N U_{\varepsilon,\alpha,n} - \sum_{n=0}^M U_{\varepsilon,\alpha,n}\right\|^2 = \sum_N^M \|U_{\varepsilon,\alpha,n}\|^2$$

(by the orthogonality of $U_{\varepsilon,\alpha,n}$ and $U_{\varepsilon,\alpha,n'}$, $n \neq n'$) and the fact that $\sum_{n=0}^\infty \|U_{\varepsilon,\alpha,n}\|^2 < \infty$, by Lemma 10.2. For any $\varepsilon > 0$ or $\varepsilon = 0$, $\alpha^2 < 4\pi$ set, in the sense of $L^2(\mu_0)$-limits, $U_{\varepsilon,\alpha} = \lim_{N\to\alpha} \sum_{n=0}^N U_{\varepsilon,\alpha,n}$.

I.10: Functional quantization

Prop. 10.2: For $\alpha^2 < 4\pi$, $(U_{\epsilon,\alpha})_{\epsilon>0}$ converges as $\epsilon \downarrow 0$ in $L^2(\mu_0)$ to $U_\alpha \equiv \lim_{N\to\infty} \sum_{n=0}^{N} U_{\alpha,n}$ (in the $L^2(\mu_0)$ − sense).

Proof: We have by construction (in the $L^2(\mu_0)$-sense)

$$U_{\epsilon,\alpha} = \lim_{N\to\infty} \sum_{n=0}^{N} U_{\epsilon,\alpha,n}$$

and

$$U_\alpha = \lim_{N\to\infty} \sum_{n=0}^{N} U_{\alpha,n}.$$

By the orthogonality of $U_{\epsilon,\alpha,n}$ in n, Prop. 10.1 and B. Levi

$$\lim_{\epsilon\downarrow 0} \|U_{\epsilon,\alpha}\|^2 = \sum_{n=0}^{\infty} \|U_{\alpha,n}\|^2.$$

On the other hand $U_\alpha = \sum_{n=0}^{\infty} U_{\alpha,n}$. From this we get $\|U_\alpha - U_{\epsilon,\alpha}\|^2 \to 0$ as $\epsilon \downarrow 0$.

Remark: In [Ku] Kusuoka has been able to prove that for $p \in (1,2)$, $\epsilon_n \equiv 2^{-2n-1}$, $\alpha^2 p < 8\pi$:

$$\sup_n \|U_{\epsilon_n,\alpha}\|_{L^p(\mu_0)} < \infty.$$

Proof of Theorem 10.1: The proof is now immediate from (10.8) and Prop. 10.2, observing that $U_\alpha \in L^2(\mu_0)$, $U_\alpha \geq 0$ μ-a.e. (since $U_{\epsilon,\alpha} \geq 0$ μ-a.e.).

∎

Corollary 10.1: For $\alpha^2 < 4\pi$ (with the limit taken in $L^2(\mu_0)$-sense)

$$\frac{d\mu_0^\alpha}{d\mu_0} = U_\alpha(\varphi) = L^2(\mu_0) - \lim_{\epsilon\downarrow 0} U_{\epsilon,\alpha}(\varphi) = \frac{d\mu_{0,\epsilon}^\alpha}{d\mu_0}(\varphi).$$

Remark:
(1) One usually introduces the natural <u>notation</u> $\int : e^{\alpha\varphi(x)} : m_g(dx)$ for $U_\alpha(\varphi)$.
(2) $:\varphi_\epsilon(x)^n:$, $:e^{\alpha\varphi_\epsilon(x)}:$ are usually called the nth Wick power resp. α-Wick exponential of the Gaussian random variable $\varphi_\epsilon(x)$; see e.g. [AHK1], [Si1].

As a consequence of this theorem $e^{-\lambda U_\alpha} \in L^\infty(\mu_0)$ for $\lambda \geq 0$. By Jensen's inequality, on the other hand, $E_{\mu_0}(e^{-\lambda U_\alpha}) \geq e^{-\lambda E_{\mu_0}(U_\alpha)} > 0$. Hence we get:

Corollary 10.2: For $\alpha^2 < 4\pi$ and $\lambda \geq 0$

$$\nu_\alpha \equiv \frac{e^{-\lambda U_\alpha}\mu_0}{\int_{\mathcal{D}'} e^{-\lambda U_\alpha(\varphi)}\mu_0(d\varphi)} \tag{10.24}$$

is a probability measure on $\mathcal{D}'$.

The measure ν_α is the generalization to a curved space, namely the surface (S, g), of the exponential interaction Euclidean measure on flat 2-dimensional spaces which has already been quoted ([AHK1]). It has been discussed in [Sc]. For other works on Euclidean quantum field theory on curved spaces, see e.g. [APS], [Wa], [DdD]. For a clearer understanding of the name "exponential interaction" one has to look closer at the multiplicative factor $e^{-\lambda U_\alpha}$ in front of the free field measure μ_0 which contains the so called "interaction term" U_α (for readers who might be unfamiliar with this type of measure (and the relatively rigorous construction) arising in Euclidean quantum field theory and classical statistical mechanics we refer to [S1], [GJ] and [AHK1,3]). Indeed for the approximation $U_{\epsilon,\alpha}$ of U_α we know the explicit form

$$U_{\epsilon,\alpha}(\varphi) = \int_S :e^{\alpha\varphi_\epsilon(x)}: m_g(dx)$$

so the name "exponential interaction" for the limit U_α, also denoted by

$$U_\alpha(\varphi) \equiv \int_S :e^{\alpha\varphi(x)}: m_g(dx),$$

becomes clear.

This can be generalized to the case of a surface S with boundary ∂S:

Corollary 10.3: Let $\tilde{m}_g$ be the measure induced on the boundary ∂S by the measure m_g on S. Then for $\alpha^2 < 4\pi$, $\lambda, \lambda' \geq 0$

$$\nu_\alpha^b \equiv \frac{e^{-U_\alpha^b}\mu_0^b}{\int_{\mathcal{D}'} e^{-U_\alpha^b(\varphi)}d\mu_0^b(d\varphi)}$$

is a probability measure on $\mathcal{D}'$, where

$$U_\alpha^b(\varphi) \equiv \lambda \int_S :e^{\alpha\varphi(x)}: m_g(dx) + \lambda' \int_{\partial S} :e^{\frac{\alpha}{2}\varphi(\eta)}: \tilde{m}_g(d\eta)$$

and $\mu_0^b(d\varphi)$ is the free field measure on S with Dirichlet boundary conditions on ∂S, defined similarly to μ_0 by (10.1) with Δ_g replaced by the Laplacian $\Delta_g^\mathcal{D}$ with Dirichlet boundary on ∂S.

Proof: The existence of U_α^b as a positive function in $L^2(\mu_0^b)$ follows, for $\alpha^2 < 4\pi$, similarly to the above, because the singularity of $(-\tilde{\Delta}_g)^{-1}$ on ∂S is only logarithmic. We then proceed as before to conclude the proof. ∎

Having in mind applications of the exponential interaction to Polyakov's bosonic strings, we now perform a small variation of the construction just presented. First of all we replace the free field measure of unit mass on (S, g) by the massless free field measure on (S, g) (but we shall keep the same notation μ_0 for it!). μ_0 is defined as the Gaussian measure on $\mathcal{D}'$ for which

$$\int_{\mathcal{D}'} e^{i<\varphi, u>} \mu_0(d\phi) = e^{-\frac{1}{2}(u, -\tilde{\Delta}_g u)} \tag{10.25}$$

where $-\tilde{\Delta}_g \equiv -\Delta_g + H$, with H the projection in $L^2(S, m_g)$ onto the subspace of constant functions on S (S being compact implies that H coincides with the projection onto the subspace of $-\Delta_g$-harmonic functions). We have $-\tilde{\Delta}_g > 0$ and $\Delta_g(\tilde{\Delta}_g)^{-1} = 1 - H$, we have (with p as in (10.3) and $H(x, y)$ denoting the kernel of H):

$$\tilde{G}(x, y) \equiv \tilde{\Delta}_g^{-1}(x, y) = \int_0^\infty e^{-t}\left(p(t, x, y) - H(x, y)\right) dt. \tag{10.26}$$

Hence all the approximants we have defined before can again be considered and the previous results, including Theorem 10.1 and its Corollary 10.1, transfer directly to cope with this massless case. In the literature, for historical reasons the massless case of the exponential model is better known under the name of "Liouville model" and it is treated in many publications at both classical and quantum levels (some indicative references are contained in [AHKPS1-2]). Therefore, in the massless case we shall call the measure ν_α of Corollary 10.2 the Liouville measure on (S, g).

For various reasons (in particular to set up relationships with string theory) this measure is not sufficient for our purposes and we need to introduce the notion of Liouville measure on (S, g) with a constant curvature term.

The construction of this new measure, which we shall denote by ν_α^c, goes as follows. For $\varphi \in \operatorname{supp} \mu_0$ one can define (by extension from smooth test functions) $<\chi_S, \varphi>$ as an $L^2(\mu_0)$-function (χ_S being the indicator function of S). For $\alpha \neq 0$ and $\alpha^2 < 4\pi$ let us set

$$V_{\alpha, \lambda}(\varphi) \equiv \lambda U_\alpha(\varphi) - \frac{1}{\alpha} <\chi_S, \varphi> \tag{10.27}$$

We remark that again by Jensen's inequality $\int_{\mathcal{D}'} e^{-V_{\alpha, \lambda}} \mu_0(d\varphi) > 0$ and define ν_α^c by

$$\nu_\alpha^c(d\varphi) \equiv \frac{e^{-V_{\alpha, \lambda}} \mu_0(d\varphi)}{\int_{\mathcal{D}'} e^{-V_{\alpha, \lambda}} \mu_0(d\varphi)}. \tag{10.28}$$

In a similar way we can also define the Liouville measure $\nu_\alpha^{c,b}$ on a Riemann surface with smooth nontrivial boundary ∂S; we shall call it Liouville measure on (S,g) with a constant curvature term and a boundary term. $\nu_\alpha^{c,b}$ is defined as in (10.28), with $V_{\alpha,\lambda}$ replaced by $V_{\alpha,\lambda}^b(\varphi) \equiv V_{\alpha,\lambda}(\varphi) - \beta <\chi_{\partial S}, \varphi>$ $+ \lambda' \int_{\partial S} : e^{\frac{\alpha}{2}\varphi} : d\eta$, where β is a real constant, $\lambda' \geq 0$ and $<\chi_{\partial S}, \varphi>$ is a well defined ν_α^c-measurable function.

Corollary 10.4: For $\lambda \geq 0$, $\alpha \neq 0$, $\alpha^2 < 4\pi$, $\nu_\alpha^c(\cdot)$ and $\nu_\alpha^{c,b}(\cdot)$ define probability measures on $\mathcal{D}'$.

Proof: Set $R(\varphi) \equiv e^{-\lambda U_\alpha(\varphi)}$ and $T(\varphi) \equiv e^{\frac{1}{\alpha}\varphi(1s)}$ and notice that

$$E_{\mu_0}(T) = e^{\frac{1}{2\alpha^2}(1s,(-\tilde{\Delta}_g)^{-1}1s)} = e^{\frac{1}{2\alpha^2}\|(-\tilde{\Delta}_g)^{-1}\|}{}_{L^1(S\times S)} < \infty,$$

S being compact and $(-\tilde{\Delta}_g)^{-1}$ being in the Hilbert–Schmidt class. Since $R \in L^\infty(\mu_0)$ it follows that $E_{\mu_0}(e^{-V_{\alpha,\lambda}}) = E_{\mu_0}(R \cdot T) < \infty$. Similarly we show $E_{\mu_0}(e^{-V_{\alpha,\lambda}^b}) < \infty$. Moreover the expectations are strictly positive, as seen by Jensen's inequality. ∎

As in [AHK1] one can show that the support of ν_α^c coincides with the support of μ_0 (regularity of ν_α^c in the sense of [AHK2]). The latter is known to consist of elements which can be tested by distributions in $H_{-1}(S)$ whereas the fields on ∂S can be tested by distributions on $H_{-\frac{1}{2}}(\partial S)$.

These results will be applied in section II.6 in relation to the discussion of bosonic strings.

I.11 Small time asymptotics for heat-kernel regularized determinants

In section I.5 we introduced and compared two standard ways to regularize naturally divergent determinants of positive elliptic operators acting on the sections of some bundle over a compact manifold. Here we want to study these divergences more carefully in the case of determinants of operators of Laplace–Beltrami type defined on a smooth boundaryless connected compact surface S, regularized by the heat-kernel method.

Let us first recall that for $\varepsilon > 0$ the heat-kernel regularized determinant of $-\Delta_g$ is by definition:

$$\text{Det}'_\varepsilon(-\Delta_g) \equiv \exp\left(-\int_\varepsilon^{+\infty} Tr(e^{t\Delta'_g})\frac{dt}{t}\right) \tag{11.1}$$

(compare with (5.17)–(5.20)), where on the right hand side Δ'_g denotes the restriction of Δ_g to the orthogonal complement of its kernel. We look at the function $F(\varepsilon) \equiv \text{Det}'_\varepsilon(-\Delta_g)$. From (11.1) it is clear that the knowledge of F depends on the knowledge of $\gamma(t) \equiv Tr(e^{t\Delta'_g})$ $\forall t \in [\varepsilon, \infty)$. However, by a trick due to Ray and Singer ([RaSi]), it is possible to show that in order to get the behaviour of $F(\varepsilon)$ for small values of ε it is sufficient to know the asymptotic behaviour of $\gamma(t)$ for $t \to 0^+$. To obtain this result we consider a one-parameter family of Laplacians.

Prop. 11.1: Let $I \subset \mathbb{R}$ and $I \ni \alpha \to g_\alpha$ be a one-parameter family of smooth Riemannian metrics on S, C^1-dependent on the parameter. Set $\Delta_\alpha \equiv \Delta_{g_\alpha}$. Then for all $t > 0$ the map $I \ni \alpha \to Tr(e^{t\Delta_\alpha})$ is C^1 and moreover we have

$$\frac{d}{d\alpha}Tr(e^{t\Delta_\alpha}) = tTr\left(\frac{d\Delta_\alpha}{d\alpha}e^{t\Delta_\alpha}\right), \tag{11.2}$$

$\frac{d\Delta_\alpha}{d\alpha}f = \frac{d}{d\alpha}(\Delta_\alpha f)$ being the strong derivative of f in $L^2(S, m_g)$, for any fixed $f \in C^2$. (Tr is here understood in the sense of the metric g_α).

Remark: In [RaSi] this proposition is in the more general context of a compact Riemannian manifold of arbitrary dimension (possibly bordered) and the operators Δ_α are Laplacians on forms.

Proof: Let us set $p_\alpha(t, x, y) \equiv e^{t\Delta_\alpha}(x, y)$ $\forall x, y \in S, t > 0$, and

$$F(\alpha, t) \equiv Tr(e^{t\Delta_\alpha}) = \int_S P_\alpha(t, x, x)\sqrt{g_\alpha}(x)dx. \tag{11.3}$$

Then for $\alpha' > \alpha$ we have

$$F(\alpha', t) - F(\alpha, t) = tTr((\Delta_{\alpha'} - \Delta_\alpha)e^{t\Delta_\alpha}) + R(\alpha', \alpha, t) \tag{11.4}$$

86   *I.11: Asymptotics for heat-kernel regularized determinants*

where $R(\alpha', \alpha, t)$ is a remainder term explicitly computable as a function of $p_\alpha, p_{\alpha'}$ and their derivatives. The main point is that by standard estimates of such kernels it can be proven that $R(\alpha', \alpha, t) = o(\alpha' - \alpha)$ so that (11.4) implies (11.2).   ∎

Corollary 11.1:   Let g_0 be a fixed Riemannian metric on S. Set $g_\alpha \equiv e^{\alpha \varphi} g_0$ for $\alpha \in [0, a]$, $a > 0$, with $\varphi \in C^\infty(S, \mathbb{R})$. Moreover set $\Delta_\alpha \equiv \Delta_{g_\alpha}$, then we have

$$-\frac{d}{d\alpha} \ln \, \mathrm{Det}'_\varepsilon(-\Delta_\alpha) = Tr(\varphi e^{\varepsilon \Delta'_\alpha}), \qquad (11.5)$$

and thus $\frac{d}{d\alpha}\Delta'_\alpha = -\varphi \Delta'_\alpha$ (since ker Δ_α does not depend on α).

Proof: Since $\frac{d\Delta_\alpha}{d\alpha} = -\varphi \Delta_\alpha$ on $C^2(S)$ and thus $\frac{d}{d\alpha}\Delta'_\alpha = -\varphi \Delta'_\alpha$ (since kerΔ_α does not depend on α) it follows from (11.2) that

$$\frac{d}{d\alpha} Tr(e^{t\Delta_\alpha}) = -t Tr(\varphi \Delta_\alpha e^{t\Delta_\alpha}) = -t \frac{d}{dt} Tr(\varphi e^{t\Delta_\alpha}). \qquad (11.6)$$

Hence (inserting primes)

$$-\frac{d}{d\alpha} \ln \, \mathrm{Det}'_\varepsilon(-\Delta_\alpha) = \int_\varepsilon^\infty \frac{d}{d\alpha} Tr(e^{t\Delta'_\alpha}) \frac{dt}{t}$$

$$= -\int_\varepsilon^\infty \frac{d}{dt} Tr(\varphi e^{t\Delta'_\alpha}) dt = Tr(\varphi e^{\varepsilon \Delta'_\alpha}).$$

∎

Remark: (11.5) is the key formula. When ε is small the asymptotic expansion of the right hand side is well-known by the work of Seeley (e.g. [See]) therefore we can exploit such an expansion and then go back to the determinant by just performing an integration.

We have specialized Corollary 11.1 to a form particularly suitable for applications to string theory and the infinite-dimensional determinants appearing there. Indeed, we can now prove the following.

Prop. 11.2:   Let S be a smooth compact connected surface. If S is boundaryless, Δ_g denotes the Laplace operator in S; if S has boundaries, Δ_g^D denotes the Laplace operator with Dirichlet boundary conditions and Δ_g^N denotes the Laplace operator with Neumann boundary conditions. Let g_0, g_1 be conformal equivalent Riemannian metrics on S, i.e. $g_1 = e^\varphi g_0$ for some $\varphi \in C^\infty(S, \mathbb{R})$. Then we have

$$\mathrm{Det}'_\varepsilon(-\Delta_{g_1}) = \left[\frac{\mathrm{Vol}_{g_1}(S)}{\mathrm{Vol}_{g_0}(S)} \exp(-W_0(\varepsilon, \varphi))\right] \mathrm{Det}'_\varepsilon(-\Delta_{g_0}) \qquad (11.7a)$$

$$\mathrm{Det}'_\varepsilon(-\Delta_{g_1}^N) = \left[\frac{\mathrm{Vol}_{g_1}(S)}{\mathrm{Vol}_{g_0}(S)} \exp(-W_0^N(\varepsilon, \varphi))\right] \mathrm{Det}'_\varepsilon(-\Delta_{g_0}^N) \qquad (11.7b)$$

I.11: Asymptotics for heat-kernel regularized determinants

$$\mathrm{Det}'_\varepsilon(-\Delta^D_{g_1}) = \left[\frac{\mathrm{Vol}_{g_1}(S)}{\mathrm{Vol}_{g_0}(S)}\exp(-W^D_0(\varepsilon,\varphi))\right]\mathrm{Det}'_\varepsilon(-\Delta^D_{g_0}) \qquad (11.7c)$$

with

$$\begin{aligned}W_0(\varepsilon,\varphi) &\equiv \frac{1}{4\pi\varepsilon}(\mathrm{Vol}_{g_1}(S) - \mathrm{Vol}_{g_0}(S)) \\ &\quad + \frac{1}{24\pi}\left(\frac{1}{2}\int_S \varphi(\eta)(-\Delta_{g_0}\varphi(\eta)\sqrt{g_0}(\eta)d\eta \right. \\ &\quad \left. + \int_S \varphi(\eta)R_{g_0}(\eta)\sqrt{g_0}(\eta)d\eta\right) + K_0(\varepsilon,\varphi)\end{aligned} \qquad (11.8)$$

$$W^N_0(\varepsilon,\varphi) = W_0(\varepsilon,\varphi) + \frac{1}{4\sqrt{\pi\varepsilon}}\int_S \sqrt{g_0}(\eta)e^{\frac{\varphi}{2}(\eta)}d\eta$$
$$\quad - \frac{1}{8\pi}\int_S \sqrt{g_0}(\eta)\partial_2\varphi(\eta)$$

and

$$W^D_0(\varepsilon,\varphi) = W_0(\varepsilon,\varphi) - \frac{1}{4\sqrt{\pi\varepsilon}}\int_S \sqrt{g_0}(\eta)e^{\frac{\varphi}{2}(\eta)}d\eta$$

where R_{g_0} is the scalar curvature with respect to g_0, and $K_0(\varepsilon,\varphi)$ is such that $K_0(\varepsilon,0) = 0$ and $\lim_{\varepsilon\to 0} K_0(\varepsilon,\varphi) = 0$.

Remark: By formula (11.7) we have fully factorized out the φ-dependence of the left hand side (for $\varphi \equiv 0$ (i.e. $g_1 = g_0$) we have $W_0(\varepsilon,0) = 0$ and (11.7) reduces to a trivial identity).

Proof: We shall only prove this proposition in the case of a boundaryless surface and we refer the reader to [D] (formulae (V.50), (V.51), (V61)) and [DOP] for further details. Keeping the notation of Corollary 11.1 we set $I \equiv [0,1]$, so $g_1 = g_{\alpha=1}$ and $g_0 = g_{\alpha=0}$. Moreover we set

$$F(\alpha) \equiv \ln \mathrm{Det}'_\varepsilon(-\Delta_\alpha).$$

Applying (11.5) and the fundamental theorem of calculus we have

$$F(0) - F(1) = \int_0^1 Tr(\varphi e^{\varepsilon\Delta'_\beta})d\beta. \qquad (11.9)$$

Since the following identity holds:

$$Tr(\varphi e^{\varepsilon\Delta_\beta}) = Vol_{g_\beta}(S))^{-1}\int_S \varphi(\eta)\sqrt{g_\beta}(\eta)d\eta + Tr(\varphi e^{\varepsilon\Delta'_\beta}),$$

we get

$$\begin{aligned}F(0) - F(1) &= \int_0^1\left(\int_S \varphi(\eta)e^{\varepsilon\Delta_\beta}(\eta,\eta)\sqrt{g_\beta}(\eta)d\eta\right)d\beta \\ &\quad - \int_0^1 (Vol_{g_\beta}(S))^{-1}\left(\int_S \varphi(\eta)\sqrt{g_\beta}(\eta)d\eta\right)d\beta.\end{aligned}$$

Since $\sqrt{g_\beta} = e^{\beta\varphi}\sqrt{g_0}$ we have

$$\int_0^1 \left((\mathrm{Vol}_{g_\beta}(S))^{-1}\int_S \varphi(\eta)\sqrt{g_\beta}(\eta)d\eta\right)d\beta = \int_0^1 \left(\frac{d}{d\beta}\ln \mathrm{Vol}_{g_\beta}(S)\right)d\beta$$
$$= \ln\frac{\mathrm{Vol}_{g_1}(S)}{\mathrm{Vol}_{g_0}(S)}.$$

Hence

$$F(1) = \ln\frac{\mathrm{Vol}_{g_1}(S)}{\mathrm{Vol}_{g_0}(S)} - \int_0^1 \left(\int_S \varphi(\eta)e^{\varepsilon\Delta_\beta}(\eta,\eta)\sqrt{g_\beta}(\eta)d\eta\right)d\beta + F(0). \quad (11.10)$$

In two dimensions the small time expansion for the heat-kernel at coincident points is (cf. [See], [G1,2])

$$e^{\varepsilon\Delta_\beta}(\eta,\eta) = \frac{1}{4\pi\varepsilon} + \frac{1}{24\pi}R_{g_\beta}(\eta) + O_\beta(\sqrt{\varepsilon})(\eta) \text{ as } \varepsilon\downarrow 0. \quad (11.11)$$

Therefore, taking into account the equality $R_{g_\beta}(\eta) = e^{-\beta\varphi(\eta)}(-\beta\Delta_{g_0}\varphi(\eta) + R_{g_0}(\eta))$, we have

$$\varphi(\eta)e^{\varepsilon\Delta_\beta}(\eta,\eta)\sqrt{g_\beta}(\eta) = \left[\frac{\varphi(\eta)e^{\beta\varphi(\eta)}}{4\pi\varepsilon} + \frac{\varphi(\eta)}{24\pi}(-\beta\Delta_{g_0}\varphi(\eta) + R_{g_0}(\eta))\right.$$
$$\left. + \varphi(\eta)e^{\beta\varphi(\eta)}O_\beta(\sqrt{\varepsilon})(\eta)\right]\sqrt{g_0}(\eta) \text{ as } \varepsilon\downarrow 0.$$
$$(11.12)$$

It follows that

$$\int_0^1 \left(\int_S \varphi(\eta)e^{\varepsilon\Delta_\beta}(\eta,\eta)\sqrt{g_\beta}(\eta)d\eta\right)d\beta =$$
$$\frac{1}{4\pi\varepsilon}\int_S \left(\int_0^1 \frac{d}{d\beta}e^{\beta\varphi(\eta)}d\beta\right)\sqrt{g_0}(\eta)d\eta \quad (11.13)$$
$$+ \frac{1}{24\pi}\int_S \varphi(\eta)\left(-\frac{1}{2}\Delta_{g_0}\varphi(\eta) + R_{g_0}(\eta)\right)\sqrt{g_0}(\eta)d\eta + K_0(\varepsilon,\varphi)$$

where

$$K_0(\varepsilon,\varphi) \equiv \int_S \left(\int_0^1 O_\beta(\sqrt{\varepsilon})(\eta)(\frac{d}{d\beta}e^{\beta\varphi(\eta)})d\beta\right)\sqrt{g_0}(\eta)d\eta. \quad (11.14)$$

Let us notice that $K_0(\varepsilon,0) = 0$ and $K_0(\varepsilon,\varphi) \to 0$ as $\varepsilon \to 0^+$. From (11.13) we obtain

$$\int_0^1 \left(\int_S \varphi(\eta)e^{\varepsilon\Delta_\beta}(\eta,\eta)\sqrt{g_\beta}(\eta)d\eta\right)d\beta = \frac{1}{4\pi\varepsilon}(\mathrm{Vol}_{g_1}(S) - \mathrm{Vol}_{g_0}(S))$$
$$+ \frac{1}{24\pi}\left(\frac{1}{2}\int_S \varphi(\eta)(-\Delta_{g_0}\varphi)(\eta)\sqrt{g_0}(\eta)d\eta + \int_S \varphi(\eta)R_{g_0}(\eta)d\eta\right) + K_0(\varepsilon,\varphi).$$
$$(11.15)$$

Inserting (11.15) into (11.10) and using the definition of $F(\alpha)$ the result (i.e. (11.7), (11.8)) follows.

∎

As a consequence of (11.7) we have also, for all $a \in \mathbb{R}$,

$$\mathrm{Det}'_\varepsilon(-\Delta_{g_1})^a = \left(\frac{\mathrm{Vol}_{g_1}(S)}{\mathrm{Vol}_{g_0}(S)}\right)^a \exp(-aW_0(\varepsilon,\varphi))\, (\,\mathrm{Det}'_\varepsilon(-\Delta_{g_0}))^a. \quad (11.16)$$

The result of Prop. 11.2 generalizes, with small modifications, to the operator $P_g^* P_g$ introduced in the previous sections (see section I.4), which is also a kind of Laplacian but acting on vector fields. Recall that $P_g^* P_g > 0$ for genus $p \geq 2$, so that $(P_g^* P_g)' = P_g^* P_g$.

Prop. 11.3: Under the hypothesis of Prop. 11.2 we have for a surface without boundary

$$\mathrm{Det}_\varepsilon(P_{g_1}^* P_{g_1}) = \left[\frac{\det H(P_{g_1}^*)}{H(P_{g_0}^*)} \exp(-W_1(\varepsilon,\varphi))\right] \mathrm{Det}_\varepsilon(P_{g_0}^* P_{g_0}) \quad (11.17)$$

with

$$\begin{aligned}W_1(\varepsilon,\varphi) \equiv\, & \frac{1}{2\pi\varepsilon}(\mathrm{Vol}_{g_1}(S) - \mathrm{Vol}_{g_0}(S)) \\ & + \frac{13}{12\pi}\left(\frac{1}{2}\int_S \varphi(\eta)(-\Delta_{g_0}\varphi)(\eta)\sqrt{g_0}(\eta)d\eta \right.\\ & \left.+ \int_S \varphi(\eta)R_{g_0}(\eta)\sqrt{g_0}(\eta)d\eta\right) \\ & + K_1(\varepsilon,\varphi),\end{aligned} \quad (11.18)$$

where $K_1(\varepsilon,0) = 0$, $K_1(\varepsilon,\varphi) \to 0$ as $\varepsilon \to 0^+$. The $H(P_{g_\alpha}^*)$, $\alpha = 0,1$, are finite-dimensional matrices built up from elements of a basis of ker $(P_{g_\alpha}^*)$, $\alpha = 0,1$, namely $H(P_g^*) = (<\Psi_g^i, \Psi_g^j>)$, where Ψ_g^i, $i = 1, \ldots, 6p-6$ is a basis of ker (P_g^*).

Remark: A similar formula holds for the case of a surface with boundary taking either Dirichlet or Neumann boundary conditions, replacing the matrices $H(P_g^*)$ by matrices $H^D(P_g^*)$ and $H^N(P_g^*)$ respectively and the term $W_1(\varepsilon,\varphi)$ by $W_2^D(\varepsilon,\varphi)$ and $W_2^N(\varepsilon,\varphi)$ respectively, where $W_1^D(\varepsilon,\varphi)$ is obtained from $W_2(\varepsilon,\varphi)$ by adding a divergent term proportional to $\frac{1}{\sqrt{\varepsilon}}$ (for $\varepsilon \downarrow 0$) and $W_1^N(\varepsilon,\varphi)$ is obtained from $W_1(\varepsilon,D)$ by adding a divergent term proportional to $\frac{1}{\sqrt{\varepsilon}}$ and a term proportional to $\int_S \sqrt{g_0}(\eta)\partial_2\varphi(\eta)d\eta$. We refer the reader to [DOP] for more details concerning this point.

Proof: This proof is entirely similar to that of Prop. 11.2. ∎

As a final result of this section we can now state the proposition which gives the complete formula for the product of regularized determinants appearing in the bosonic string measure with the so called conformal factor, or "Liouville field", removed.

Prop. 11.4:
(1) Let S be a compact boundaryless Riemannian surface. Let g_1, g_0 be conformal metrics on S, i.e. $g_1 = e^\varphi g_0$, $\varphi \in C^\infty(S, \mathbb{R})$. For $\varepsilon > 0$ we have, with the notation of the previous propositions,

$$\left(\frac{\text{Det}'_\varepsilon(-\Delta_{g_1})}{\text{Vol}_{g_1}(S)}\right)^{-\frac{d}{2}} (\text{Det}_\varepsilon(P^*_{g_1} P_{g_1}))^{\frac{1}{2}}$$
$$= \left(\frac{\det H(P^*_{g_1})}{\det H(P^*_{g_0})}\right)^{\frac{1}{2}} \exp(F_d(\varepsilon, \varphi)) \left(\frac{\text{Det}'_\varepsilon(-\Delta_{g_0})}{\text{Vol}_{g_0}(S)}\right)^{-\frac{d}{2}} (\text{Det}_\varepsilon(P^*_{g_0} P_{g_0}))^{\frac{1}{2}}$$
(11.19)

with

$$F_d(\varepsilon, \varphi) \equiv \frac{d}{2} W_0(\varepsilon, \varphi) - \frac{1}{2} W_1(\varepsilon, \varphi) = \left(\frac{d-2}{8\pi\varepsilon}\right) (\text{Vol}_{g_1}(S) - \text{Vol}_{g_0}(S))$$
$$+ \left(\frac{d-26}{48\pi}\right) \left(\frac{1}{2} \int_S \varphi(\eta)(-\Delta_{g_0})(\eta) \sqrt{g_0} \varphi(\eta) d\eta\right.$$
$$+ \int_S \varphi(\eta) R_{g_0}(\eta) \sqrt{g_0}(\eta) d\eta \right)$$
$$+ K(\varepsilon, \varphi)$$
(11.20)

where $K(\varepsilon, \varphi) \to 0$ as $\varepsilon \to 0^+$.

(2) For a Riemannian surface S with boundary, and Dirichlet or Neumann conditions, the asymptotic formula (11.20) generalizes up to an additional divergent term in $\frac{1}{\sqrt{\varepsilon}}$ and heuristic integrals on the boundary involving a linear term in φ and an exponential term in $\varphi/2$.

Remark: In (11.19), (11.20) d is a natural number, interpreted in the context of string theory as the dimension of the target space where strings are embedded (see chapter II).

Proof: It follows directly from Propositions 11.2 and 11.3, using (11.16). ∎

I.11: Asymptotics for heat-kernel regularized determinants

Remark: As we shall see in section II.8, in the construction of the Polyakov measure, one can compensate the additional divergent term proportional to $\frac{1}{\sqrt{\epsilon}}$ arising in the asymptotic expansion for surfaces with boundary by introducing an additional renormalizing term in the action, so that in the case of Dirichlet boundary conditions the same procedure as for boundaryless surfaces can be applied. The case of Neumann boundary conditions is however different because of the term in the heat-kernel expansion proportional to $\int_S \sqrt{g_0}(\eta)\partial_2\varphi(\eta)d\eta$ which cannot be compensated without an important modification of the original Polyakov action.

II.1 Quantization by functional integrals

Since the work of Feynman (see e.g. [AHK] and references therein) one of the most heuristically appealing methods of quantization of a given classical theory consists of writing a heuristic infinite dimensional measure, expressed essentially in terms of the action integral associated with the classical theory, and a heuristic "infinite-dimensional Riemann–Lebesgue volume measure". The type of space on which the measure is to be considered depends on the problem at hand, but usually, in the physical literature, attention is paid to boundary conditions or symmetry considerations, rather than precise considerations of its functional-analytic properties.

So e.g. the quantization by functional integrals of a classical mechanical particle moving in $I\!R^d$ under the influence of a continuous potential $V(x)$, described quantum-mechanically by the Schrödinger equation

$$i\frac{\partial}{\partial t}\psi = -\frac{1}{2}\Delta\psi + V\psi \tag{1.1}$$

for $\psi : I\!R_+ \times I\!R^d \to \mathbb{C}$, with initial condition $\psi(0,x) \equiv \varphi(x)$, φ a bounded continuous complex function $I\!R^d$, is done by building a heuristic (Feynman) complex measure of the form

$$\mu_F(d\gamma) = \frac{1}{Z}e^{iS_t(\gamma)}d[\gamma], \tag{1.2}$$

with Z a heuristic normalization s.t. $\int \mu_F(d\gamma) = 1$,

$$S_t(\gamma) = \frac{1}{2}\int_0^t |\dot\gamma(s)|^2\,ds - \int_0^t V(\gamma(s))ds,$$

the classical action along a path $\gamma : [0,t] \to I\!R^d$, s.t. $\gamma(t) = x$, $d[\gamma]$ being a heuristic Lebesgue measure $\prod_{\tau \in [0,t]} d\gamma(\tau)$ on the space Γ^x of "all paths" ending at point x at time t (with $d\gamma(\tau)$ a Lebesgue measure on the τ-copy of $I\!R^d$). Γ^x is supposed to be a certain subspace of $x + I\!R^{[0,t]}$. Heuristically one has

$$\psi(t,x) = \int \varphi(\gamma(0))\mu_F(d\gamma). \tag{1.3}$$

There are several well-known solutions of the problem of giving a suitable mathematical realization of such a heuristic measure and of the corresponding representation of the solutions of the Schrödinger equation. A first distinction between them is whether or not they involve some analytic continuation (in t or V or other parameters, hidden in our notation: in fact in our equation

II.1: Quantization by functional integrals

we have assumed units s.t. both Planck's constant h and the mass m of the particle are 1). Without analytic continuation two basic methods are known: a sequential one, based on the Lie–Trotter formula (see [Ne] and e.g. [AHK]), and one based on Parseval's formula for Fourier transforms which has its origins in work by K. Ito [I] (see references in [AHK], see also [AB]). Especially in connection with the corresponding quantization problem of quantum fields, a method which corresponds to analytic continuation in t, more precisely to replacing $(t, t \geq 0)$ by $(-it, t \geq 0)$, turns out to be useful. Heuristically such a substitution transforms the Schrödinger equation (1.1) into a heat equation with a "source" V:

$$\begin{cases} \frac{\partial}{\partial t}\psi = \frac{1}{2}\Delta\psi - V\psi, \\ \psi(0, x) = \varphi(x), \end{cases} \tag{1.4}$$

the heuristic "Feynman measure" into the heuristic "(Euclidean) probability measure" of the form

$$\mu_E(d\gamma) = \frac{1}{Z} e^{-S_t(\gamma)} d[\gamma], \tag{1.5}$$

Z being again a "normalization constant" such that heuristically

$$\int \mu_E(d\gamma) = 1.$$

(1.3) is then replaced by

$$\psi(t, x) = \int \varphi(\gamma(0) + x) \mu_E(d\gamma). \tag{1.6}$$

As remarked originally by Kac, μ_E can be given a mathematical meaning, at least under some restrictions on V (e.g. V bounded continuous) as a measure on Wiener space $W \equiv C_0([0, t]; \mathbb{R}^d)$ of continuous paths starting at 0 at time 0, absolutely continuous with respect to Wiener measure μ_W, s.t.

$$\mu_E(d\gamma) \equiv Z^{-1} \int_W e^{-\int_0^t V(\gamma(s)+x)} d\mu_W(\gamma(t) + x)$$

with $Z \equiv \int_W e^{-\int_0^t (\gamma(s)+x)} d\mu_W(\gamma(t)+x)$. (1.6) is then the expectation of the function $\gamma \to \varphi(\gamma(0) + x)$ on W with respect to μ_E.

The paths γ in W are then the paths of a Wiener process (Brownian motion) in $\mathbb{R}^d$. The recovery of solutions of (1.1) from (1.6) is then a theorem about analytic continuation of $t \to \psi(t, \cdot)$ as function of t from $(t, t \geq 0)$ to $(it, t \geq 0)$, see e.g. [APS]. Analogous formulae and procedures can be derived to represent observables and other quantum mechanical quantities; see e.g. [AHK], [S3].

In these terms, the problem of "quantization by functional integration" then consists in constructing mathematically a probability measure which heuristically is of the form (1.5). Similar problems and formulations can be found in the problems of the quantization of classical fields. Let us consider for simplicity a scalar classical relativistic field φ over $\mathbb{R}^{d-1}, d \geq 1$. Its action in a space-time rectangular domain $\Lambda \equiv [0,t] \times [a_i, b_i], i = 1, \ldots, d-1$ is given by

$$S_\Lambda(\varphi) = \frac{1}{2} \int_\Lambda \left|\frac{\partial \varphi}{\partial s}(s,x)\right|^2 dsdx - \frac{1}{2} \sum_{i=1}^d \int_\Lambda \left|\frac{\partial}{\partial x_i}\varphi(s,x)\right|^2 dsdx +$$
$$\int_\Lambda w\left(\varphi(s,x)\right) dsdx,$$

with $s \in \mathbb{R}$, $x \in \mathbb{R}^{d-1}$.

w is the total "interaction density", usually of the form $w(\lambda) = \frac{1}{2}m^2\lambda^2 + v(\lambda)$ with $m^2 \geq 0$ a constant (mass) and v nonlinear, nonquadratic function, the proper interaction. The heuristic Feynman measure is again of the type

$$\mu_F(d\varphi) = \frac{1}{Z} e^{iS_\Lambda(\varphi)} d[\varphi]$$

with $\varphi : \Lambda \to \mathbb{R}$ s.t. $\varphi(0,x) = \varphi_0(x)$, with φ_0 a given "boundary condition". $d[\varphi]$ is a heuristic Lebesgue measure $\Pi_{(s,x)\in\Lambda} d\varphi(s,x)$ on the space Γ consisting of all paths of the form $\varphi_0 + \mathbb{R}^\Lambda$ ($\mathbb{R}^\Lambda$ being the space of "all maps" from Λ into $\mathbb{R}$).

A rigorous mathematical construction of a measure related to this measure for $v = 0$ (or for $v \neq 0$ but with a "regularization") (and $m > 0$ for $d = 1, 2$) is contained in [AHK]. The corresponding heuristic measure obtained by analytic continuation $t \to -it$ is given by

$$\mu_E(d\varphi) = \frac{1}{Z} e^{-S_\Lambda^E(\varphi)} d[\varphi]$$

with

$$S_\Lambda^E(\varphi) = \frac{1}{2} \int_\Lambda \left|\frac{\partial \varphi(s,x)}{\partial s}\right|^2 dsdx + \frac{1}{2} \sum_{i=1}^d \int_\Lambda \left|\frac{\partial}{\partial x_i}\varphi(s,x)\right|^2 dsdx +$$
$$\int_\Lambda w\left(\varphi(s,x)\right) dsdx.$$

The rigorous mathematical construction of this measure as a probability measure on $\mathcal{S}'(\mathbb{R}^d)$ has been given in several situations.

For $v = 0$, (and $m > 0$ for $d = 1, 2$), μ_E is realized as the measure associated with Nelson's free field; see section I.9.

II.1: Quantization by functional integrals

For $d = 1$, $m = 0$ one has a measure obtained from Wiener measure by perturbing by a Feynman-Kac functional given by the function w.

For $d = 2$, $m > 0$ and suitable v in a certain analytic class J there exists a modification, "renormalization", of the term $\int_\Lambda v((\varphi(s,x))\,dsdx$ which preserves the good formal properties of S_Λ^E and enables one to interpret the corresponding modified measure $\tilde{\mu}_E$ as a probability measure on $\mathcal{S}'(\mathbb{R}^2)$, given by the "interaction v". J contains certain polynomials, trigonometric and exponential functions; see e.g. [GJ], [S1], [AHK1], [AFHKL], [AHK3]. The recovery of physically intereresting quantities of the relativistic quantum fields associated with the classical field described from the above action by analytic continuation $t \to -it$ can be carried through in these cases. In this sense, one constructs a relativistic quantized field theory via the construction of a probability measure on a space $(\mathcal{S}'(\mathbb{R}^d))$ of field configurations (φ).

Similar constructions can be done, heuristically, for vector-valued fields and gauge fields; see e.g. [GrKS], [AHKH], [AHKHK], [AIK], [AS] for corresponding mathematical results, basically for $d = 2, 4$.

In all these situations the basic mathematical problem thus consists in constructing a certain probability measure μ of the heuristic form

$$\mu(dX) = \frac{1}{Z} e^{-S(X)} d[X],$$

with S an action functional over a space ("path space") of fields (or processes) X. We shall see in the next section how Polyakov's approach to the quantization of classical relativistic strings proceeds along similar lines.

II.2 The Polyakov measure

In the previous section, we recalled the main steps of the functional Euclidean quantization for classical and point-particle field theories. In this section, we want to apply this procedure to bosonic strings, thus quantizing the classical action given by (I.1.4) [†]. We shall give a heuristic description of the main steps of this quantization procedure and then in further sections investigate each of these steps precisely within a mathematical framework.

Let us first describe the classical bosonic string. A direct generalization of the classical action for the free point particles given by the length of the path described by the particle would lead to the action (I.1.1) which corresponds to the area of the surface described by the moving string. This action for the free string is clearly not a quadratic function of X and hence, when quantized, gives rise to a formal measure which is not Gaussian and therefore difficult to handle by the usual techniques of functional integration. However, on introducing an extra parameter – namely the metric g – and replacing the action (I.1.1) by (I.1.4), the minimization of the action yields the same classical equations of motion (I.1.2) on one hand and (I.1.5) combined with (I.1.6) on the other hand.

Notice that unlike the action $A(X(\Sigma))$ given by (I.1.1), the action $D(X, g)$ given by (I.1.4) is quadratic in X and can be written in the following form:

$$D(X, g) = -\frac{1}{2} \int_S \sqrt{\det g}\, \Delta_g X^\alpha X^\alpha \, dz^1 dz^2, \qquad (2.1)$$

where Δ_g is the Laplace–Beltrami operator given by (I.4.7). From now on, we work with this two-parameter dependent action $D(X, g)$.

The functional quantization procedure then requires building a measure on the space $\mathcal{F} \times \mathcal{M}$ where $\mathcal{M}$ is the space of smooth Riemannian metrics (see (I.2.6)) and $\mathcal{F}$ the space of smooth embeddings of the surface S into $\mathbb{R}^d$, which, in this section, will always be considered of topological genus $p > 1$. Following the general functional quantization procedure described in the former section, we thus want to build a formal measure

$$\mu_E(dx, dg) \equiv \frac{1}{Z} e^{-D(X,g)} d_g[X]\, d[g] \qquad (2.2)$$

with the corresponding partition function

$$Z = \int_{\mathcal{F} \times \mathcal{M}} e^{-D(X,g)} d_g[X]\, d[g] \qquad (2.3)$$

[†] In this chapter formula (I.x) refers to formula x of Chapter I, formula (y) to formula y of Chapter II.

II.2: The Polyakov measure

where $d_g[X]$ and $d[g]$ are formal Lebesgue measures on $\mathcal{F}$ and $\mathcal{M}$. The measure on $\mathcal{F}$ is g-dependent because the underlying scalar product $(\cdot,\cdot)_g$ on the tangent space to $\mathcal{F}$ is g-dependent (see section I.2).

Let us comment about the gauge invariance of the theory. The classical action $D(X,g)$ is invariant under the diffeomorphism group $\mathcal{D}$ (see section I.2) and under the action of the conformal group C defined in section I.3 since multiplying g by a function e^φ does not change the value of $D(X,g)$.

However, the formal Lebesgue measures are no longer invariant under the action of the conformal group C, in the following sense. They are formally infinite products of Lebesgue measures on $I\!R$, namely $d_g[X] = \prod_{n \in I\!N} dX_n$, $d[g] = \prod_{n \in I\!N} dg_n$, where $(X_n)_{n \in I\!N}$ (resp. (g_n)) are the components of X (resp. g) along an orthonormal basis of $L^2(S, I\!R^d)$ (resp. $L^2(S^2T^*)$) and where $L^2(S, I\!R^d)$ is the closure of $C^\infty(S, I\!R^d)$ w.r.t. the scalar product (I.2.3) (resp. $L^2(S^2T^*)$ is the closure of the space of smooth symmetric covariant 2-tensors – on which the manifold $\mathcal{M}$ is modeled – w.r.t. the scalar products (I.2.10)). Both the scalar products (I.2.3) and (I.2.10) are only invariant under the action of diffeomorphisms, not under the action of the conformal group (see Lemmas I.2.1 and I.2.3), and the corresponding formal Lebesgue measures $d_g[X]$ and $d[g]$ thus have these same features.

Notice that the invariance under the diffeomorphism group of both the classical action $D(X,g)$ and the formal measures $d_g[X]\,d[g]$ leads to an overcounting when computing expectations of diffeomorphism invariant functionals $\mathcal{F}(X,g)$ on $\mathcal{F} \times \mathcal{M}$. However, this overcounting is formally compensated by the partition function Z which contains the "volume" of the diffeomorphism group remaining from this invariance.

Until section II.8, we shall assume that the surface S is boundaryless. The case of a surface with boundary will be looked upon more closely in section II.8 (for this we also refer the reader to [A], [Jas2]).

Step 1: The integration on the space of embeddings

The formal characteristic function of the measure $d_g[X]$ reads

$$\chi_g(Y) \equiv \int_\mathcal{F} e^{i(X,Y)_g}\, e^{-D(X,g)}\, d_g[X] \tag{2.4}$$

and we have by a heuristic Gaussian integration with respect to X

$$\frac{1}{Z}\int_\mathcal{M} \chi_g(Y)d[g] = \frac{1}{Z}\int_\mathcal{M} e^{-\frac{1}{2}(Y,Y)_g} \left(\frac{\text{``det}'\text{''}(-\Delta_g)}{\int_S \sqrt{\det g(z)}dz}\right)^{-\frac{d}{2}} d[g] \tag{2.5}$$

where "det$'$" is a formal determinant which arises as a result of the Gaussian integration with respect to the formal measure $e^{-\frac{1}{2}(\Delta_g X,X)_g}d_g[X]$ on the embedding space.

As we saw in section I.5, these determinants are obtained as limits of heat-kernel regularized determinants "det $h_\varepsilon(-\Delta_g)$" when ε goes to zero after removal of the divergent terms.

The removal of divergent terms is taken care of by introducing a regularization, parametrized by $\varepsilon > 0$, and compensating terms in the functional measure, namely in the case of strings by introducing a "cosmological term" $e^{-\mu_\varepsilon^2 \int_S \sqrt{\det g}\, dz}$, where $\mu_\varepsilon^2 \to +\infty$, when $\varepsilon \downarrow 0$, and this divergence cancels the divergent terms in the determinants, as $\varepsilon \downarrow 0$.

We therefore introduce the heuristic measure

$$d\nu[g] \equiv \frac{1}{Z} e^{-\mu^2 \int_S \sqrt{\det g}\, dz} Z_g\, d[g] = n(g)\, d[g] \qquad (2.6)$$

where

$$Z_g \equiv \int_\mathcal{F} e^{-D(X,g)} d_g[X] = \chi_g(0) = \left(\frac{\text{"det'"}(-\Delta_g)}{\int_S \sqrt{\det g}\, dz} \right)^{-\frac{d}{2}},$$

$$n(g) \equiv \frac{1}{Z} e^{-\mu^2 \int_S \sqrt{\det g}\, dz} Z_g.$$

Step 2: The Faddeev–Popov procedure

A stochastic interpretation of the Faddeev–Popov procedure, by which projections of formal Lebesgue measures defined on the total (infinite-dimensional) space of a principal fibre bundle onto the quotient space are interpreted in terms of projections of well-defined regularised Brownian motions, is discussed in [AP1,2]. Here, we shall only consider formal expectation values. Expectation values with respect to the formal measure $d\nu[g]$ are of the form

$$<\mathcal{F}> \equiv \int_\mathcal{M} \mathcal{F}(g)\, d[g] \qquad (2.7)$$

where

$$\mathcal{F}(g) = \frac{1}{Z} e^{-\mu^2 \int_S \sqrt{\det g}\, dz} F(g)\, Z_g$$

for some real-valued function F on $\mathcal{M}$. From the result of section I.3, we know that the manifold $\mathcal{M}$ is a C^∞ principal trivial fibre bundle over the Teichmüller space T_p of the surface S with genus $p > 1$ with structure group $G = D_0\ C$, i.e., with the notation of section I.3:

$$\mathcal{M} \to \mathcal{M}/D_0 \odot C \simeq T_p, \qquad (2.8)$$

so that a metric $g \in \mathcal{M}$ can be parametrized by $(t,f,\varphi) \in T_p \times D_0 \times C$. Transforming the formal integral (2.7) into a formal integral on $T_p \times D_0 \times C$ yields

$$\int_\mathcal{M} \mathcal{F}(g)\, d[g] = \int_{T_p \times D_0 \times C} \mathcal{F}(t,f,\varphi) \text{"}(\det F_g^* F_g)^{\frac{1}{2}}\text{"}\, dt\, d[f]\, d[\varphi] \qquad (2.9)$$

II.2: The Polyakov measure

where F_g is the Faddeev–Popov operator tangent to the map

$$T_p \times D_0 \times C \to \mathcal{M}$$
$$(t, f, \varphi) \to g = f^* e^\varphi g_t,$$

"$(\det F_g^* F_g)^{\frac{1}{2}}$" is its formal determinant and $\mathcal{F}(t, f, \varphi)$ is the expression of $\mathcal{F}(g)$ in terms of the new parameters (t, f, φ). Here as before dt, $d[f]$ and $d[\varphi]$ are formal Lebesgue measures on T_p, D_0 and C respectively.

If F is invariant under the action of the group D_0, then the integrand in (2.7) is invariant under the action of the group D_0 and we can formally integrate out the volume of the group D_0 (the partition function Z contains the volume of D which formally compensates the volume arising in the numerator, this compensation giving rise to a constant which merely depends on the genus of the surface, namely the volume of the mapping class group Γ_p, p being the genus) so that we shall in fact work with the formal measure

$$d\mu[\varphi, t] = n(g) \cdot (\text{``}\det\text{''} F_g^* F_g)^{\frac{1}{2}} dt\, d[\varphi]. \tag{2.10}$$

Step 3: The Polyakov measure in noncritical dimensions and the Liouville measure

Applying the results of section I.5 and section I.7 to the operators $-\Delta_g$ and $P_g^* P_g$ where P_g was defined in (I.4.9), we can compute the asymptotics of the heat-kernel regularized determinants $\mathrm{Det}'_\varepsilon(-\Delta_g)$ and $\mathrm{Det}_\varepsilon(P_g^* P_g)$ when ε goes to zero. After introducing a well-chosen cosmological term $e^{-\mu_\varepsilon^2 \int_S \sqrt{\det g}\, dz}$ which cancels the divergences arising in the expression

$$\left(\frac{\mathrm{Det}'_\varepsilon(-\Delta'_g)}{\int_S \sqrt{\det g(z)}\, dz} \right)^{-\frac{d}{2}} \cdot \mathrm{Det}_\varepsilon(P_g^* P_g)^{\frac{1}{2}},$$

we obtain a renormalized expression, μ_{ren}, of the formal measure μ in (2.10):

$$d\mu_{\mathrm{ren}}[\psi, t] = e^{-\mathcal{L}(\psi, t)} h(t) d[\psi] dt \tag{2.11}$$

where

$$\mathcal{L}(\psi, t) \equiv \left[\frac{1}{2} \int_S \sqrt{\det(g_t)} (-(\Delta_{g_t} \psi)(z))\, \psi(z) dz + \lambda \int_S \sqrt{\det g_t}\, e^{\alpha \psi}(z) dz + \frac{1}{2} \int_S \sqrt{\det g_t}\, \psi dz \right],$$

$h(t)$ being a formal density on T_p, α, λ are two constants, the constant α depending on the dimension d of the Euclidean space $\mathbb{R}^d$ in which the string is embedded, and $\{g_t\}$ is a family of Riemannian metrics indexed by $t \in T_p$ with curvature -1. (We remark that a heuristic Liouville measure with boundary is discussed in [Jas4].)

When $d \leq 13$, the formal measure $e^{-\mathcal{L}(\psi,t)}d[\psi]$ indexed by t arising in (2.11) can then be given a precise meaning by using the Liouville measure described in section I.10.

Notice that the formal density $e^{-\mathcal{L}(\psi,t)}$ becomes identically 1 in critical dimension 26, and we are left with a formal measure on the Teichmüller space T_p, namely $h(t)dt$.

Step 4: The Polyakov measure in critical dimension

A priori the formal density $h(t)$ is not a function on T_p and, as we shall see in section II.7, there are anomalies which prevent us from defining a measure of the form $h(t) dt$ on the Teichmüller space. However, in critical dimension $d = 26$, these anomalies disappear and the Polyakov measure can be described precisely as a measure on the moduli space $M_p = \mathcal{M}/D \odot C$.

Step 5: Correlation functions

When the surface S has boundaries $\Gamma_1, \Gamma_2, \ldots, \Gamma_k$ an analogous reduction of the Polyakov model to the Liouville model can be carried out (see also [A], [Jas3]). As a function of two boundary curves Γ_1 and Γ_2, the partition function (Γ_1, Γ_2) can then be interpreted as a correlation function, namely the formal probability for the string to go from Γ_1 to Γ_2. We can generalize the above discussion by writing

$$Z[\Gamma_1, \Gamma_2] = \int_C \int_{T_p} j(\Gamma_1, \Gamma_2) e^{-\mathcal{L}_b(\varphi,t)} h(t) dt \, d[\varphi]$$

where $j(\Gamma_1, \Gamma_2)$ depends only on t and $e^{-\mathcal{L}_b(\varphi,t)}d[\varphi]$ is the formal measure corresponding to a Liouville model with boundary.

These main steps in the functional quantization of bosonic strings will be described precisely in the following sections in this same order.

II.3 Formal Lebesgue measures on Hilbert spaces

In this section, we give a precise description of the rules we shall need in further sections to compute formal integrals on Hilbert spaces w.r.t. formal Lebesgue measures. We shall in particular compute formal Gaussian integrals on Hilbert spaces which naturally arise in the functional quantization procedure.

Let H be a separable Hilbert space equipped with a scalar product $<\cdot,\cdot>$ and let $(e_n)_{n\in\mathbb{N}}$ be a fixed orthonormal basis of H. Our aim in this chapter is to describe rules for operating with the formal Lebesgue measure $d[h] \equiv \prod_n dh_n$ on H relative to the orthonormal basis $\mathcal{E} \equiv (e_n)$, $e_n \in H$ where $h = \sum_n h_n e_n$. We shall also give transformation rules for expressions of the form $\int_H \mathcal{F}(h)d[h]$ where $\mathcal{F}$ is a functional defined on H.
For a finite subset $I \subset \mathbb{N}$, $I = \{i_1,\ldots,i_N\}$, we define the map

$$\phi_I^{\mathcal{E}} : \mathbb{R}^N \to H$$
$$(x_{i_1},\ldots,x_{i_N}) \to \sum_{i\in I} x_i e_i. \tag{3.1}$$

For $i \in I$, we shall simply set $\phi_i^{\mathcal{E}} = \phi_{\{i\}}^{\mathcal{E}}$. To simplify the notation, we shall drop the index $\mathcal{E}$ which specifies the orthonormal basis.
For $I \subset \mathbb{N}$, let P_I denote the orthogonal projection onto the vector space spanned by $e_i, i \in I$.
Let us now introduce a class $\mathcal{C}(H)$ of suitable Borel functionals.

Definition 3.1: A functional $\mathcal{F}$ on H lies in $\mathcal{C}(H)$ if and only if
(1) $\mathcal{F} \circ \phi_I = \prod_{i\in I} \mathcal{F} \circ \phi_i \quad \forall I \subset \mathbb{N}$, I finite
(2) $\mathcal{F} \circ \phi_i \in \mathcal{B}(\mathbb{R})$ for all $i \in \mathbb{N}$
Here $\mathcal{B}(\mathbb{R})$ denotes the Borel functions on $\mathbb{R}$.

Example 3.1: Let A be a self-adjoint operator on H with pure point spectrum $(\lambda_n)_{n\in\mathbb{N}}$. We choose the basis $\mathcal{E} = (e_n)_{n\in\mathbb{N}}$ to be an orthormal basis of eigenvectors for A; then the functional defined for $h \in H$ by $\mathcal{F}_0(h) \equiv e^{-\frac{1}{2}<Ah,h>}$ lies in $\mathcal{C}(H)$ since we have $\mathcal{F}_0(h) = \prod_{n\in\mathbb{N}} e^{-\frac{1}{2}\lambda_n h_n^2}$.

Lemma 3.1: Let $\mathcal{F} \in \mathcal{C}(H)$ such that $\sum_{n\in\mathbb{N}} |\frac{1}{\sqrt{2\pi}} \int_{\mathbb{R}} \mathcal{F} \circ \phi_n(x)dx - 1| < \infty$. Then the limit when the finite subset I tends to $\mathbb{N}$ of the integrals $(\frac{1}{2\pi})^{\frac{|I|}{2}} \int_{\mathbb{R}^{|I|}} \mathcal{F} \circ \phi_I(x_{i_1},\ldots,x_{i_N})dx_{i_1}\ldots dx_{i_N}$ is well defined, independent of the sequence of subsets tending to $\mathbb{N}$ and different from 0.

Proof: Since $\mathcal{F} \in \mathcal{C}(H)$, we have for every finite set I

$$\left(\frac{1}{2\pi}\right)^{\frac{|I|}{2}} \int_{\mathbb{R}^{|I|}} \mathcal{F} \circ \phi_I(x_{i_1}, \ldots, x_{i_N}) dx_{i_1} \ldots dx_{i_n} = \prod_{i \in I} \left(\frac{1}{\sqrt{2\pi}} \int_{\mathbb{R}} \mathcal{F} \circ \phi_i(x) dx\right)$$

and the latter product converges when $I \to \mathbb{N}$ whenever the series

$$\sum_n \left|\left(\frac{1}{2\pi}\right)^{\frac{1}{2}} \int_{\mathbb{R}} \mathcal{F} \circ \phi_n(x) dx - 1\right|$$

converges. ∎

Let us now introduce a class of integrable functions.
Definition 3.2: We set

$$\mathcal{L}^1_{\mathcal{E}}(H) \equiv \{\mathcal{F} \in \mathcal{C}(H), \sum_{n \in \mathbb{N}} \left|\frac{1}{2\pi} \int_{\mathbb{R}} \mathcal{F} \circ \phi_n(x) dx - 1\right| < \infty\},$$

and for $\mathcal{F} \in \mathcal{L}^1_{\mathcal{E}}(H)$, we set

$$\int_H \mathcal{F}(h) d[h] \equiv \lim_{I \to \mathbb{N}} (\frac{1}{2\pi})^{\frac{|I|}{2}} \int_{\mathbb{R}^{|I|}} \mathcal{F} \circ \phi_I(x_{i_1}, \ldots, x_{i_{|I|}}) dx_{i_1} \ldots dx_{i_{|I|}}$$

In the following, we shall drop the index $\mathcal{E}$ in $\mathcal{L}^1_{\mathcal{E}}(H)$.

In the following lemma, we give an example of an $\mathcal{L}^1$ function on H which will enter in formal Gaussian integrals on Hilbert spaces.

Lemma 3.2: Let A be a strictly positive self-adjoint operator on H with purely discrete spectrum. The functional defined on H by $\mathcal{F}_0(h) \equiv e^{-\frac{1}{2}<Ah,h>}$ lies in $\mathcal{L}^1(H)$ if and only if the operator A is of the form "$\mathbb{1}$ + trace class operator" and

$$\int_H \mathcal{F}_0(h) d[h] = \text{Det}(A)^{-\frac{1}{2}}.$$

Proof: The fact that $\mathcal{F}_0$ lies in $\mathcal{C}(H)$ was checked in Example 3.1. Setting $a_n \equiv \frac{1}{2\pi} \int_{\mathbb{R}} e^{-\frac{1}{2}\lambda_n x^2} dx$ we have $a_n = \lambda_n^{-\frac{1}{2}}$. The operator A is of the form "$\mathbb{1}$ + trace class" if and only if we can write $\lambda_n = \mu_n + 1$ with $\sum_n |\mu_n| < \infty$. The general term of the series $\sum_n |a_n - 1|$ being equivalent to $\frac{1}{2}|\mu_n|$, the series is convergent whenever the operator A is of the form "$\mathbb{1}$+ a trace class operator".The computation of the integral then follows from Gaussian

II.3: Formal Lebesgue measures on Hilbert spaces

integral computations in $\mathbb{R}$ using the fact that for every eigenvalue λ_n of A, we have

$$\left(\frac{1}{2\pi}\right)^{\frac{1}{2}} \int_{\mathbb{R}} e^{-\frac{1}{2}\lambda_n x^2} dx = \lambda_n^{-\frac{1}{2}}.$$

∎

We now investigate transformations of formal Lebesgue measures in Hilbert spaces. Such transformations arise naturally in the context of gauge theories and in particular in the Faddeev–Popov procedure for strings.

Lemma 3.3: Let A be a one-to-one self-adjoint operator on H with purely discrete spectrum. The basis $\mathcal{E}$ in H is chosen as an orthonormal basis of eigenvectors of A. Let $\mathcal{F}$ be a functional in $\mathcal{C}(H)$. Then the functional $\mathcal{F} \circ A$ lies in $\mathcal{C}(H)$. If moreover $\mathcal{F}$ is in $\mathcal{L}^1(H)$, then $\mathcal{F} \circ A$ is in $\mathcal{L}^1(H)$ if and only if A is of the form "$\mathbb{1}$ + trace class operator".

Proof: Let us denote by $(\lambda_n)_{n \in \mathbb{N}}$ the spectrum of A. The fact that $\mathcal{F} \circ A$ lies in $\mathcal{C}(H)$ easily follows from the fact that $A \circ \phi_n(x) = \phi_n(\lambda_n \cdot x)$. Moreover

$$\sum_n \left| \frac{1}{2\pi} \int_{\mathbb{R}} \mathcal{F} \circ A \circ \phi_n(x) dx - 1 \right| < \infty$$

$$\Leftrightarrow \sum_n \left| \frac{1}{2\pi} \int_{\mathbb{R}} \mathcal{F} \circ \phi_n(\lambda_n x) dx - 1 \right| < \infty$$

$$\Leftrightarrow \sum_n \left| \lambda_n^{-1} \frac{1}{2\pi} \int_{\mathbb{R}} \mathcal{F} \circ \phi_n(x) dx - 1 \right| < \infty$$

$$\Leftrightarrow \left| \left(\prod_n \lambda_n^{-1}\right) \prod_n \frac{1}{2\pi} \int_{\mathbb{R}} \mathcal{F} \circ \phi_n(x) dx \right| < \infty$$

which shows that if $\mathcal{F} \in \mathcal{L}^1(H)$, then $\mathcal{F} \circ A$ lies in $\mathcal{L}^1(H)$ whenever $\prod_n \lambda_n$ is finite, i.e. whenever A is of the form "$\mathbb{1}$ + trace class". ∎

Let now H and H_1 be two separable Hilbert spaces and B be an injective operator from H onto H_1. We shall assume that the operator B^*B has pure point spectrum and we choose $\mathcal{E} = (e_n)_{n \in \mathbb{N}}$ to be an orthonormal basis of eigenvectors of B^*B associated with the eigenvalues $(\lambda_n)_{n \in \mathbb{N}}$. The space H_1 is equipped with a natural orthonormal basis $\mathcal{E}_1 \equiv (e_n^1)_{n \in \mathbb{N}}$, namely

$$e_n^1 \equiv \frac{Be_n}{\sqrt{\lambda_n}}.$$

In the same way as for H, we define on H_1 the sets $\mathcal{C}(H_1)$ and $\mathcal{L}^1(H_1)$ relative to the basis $\mathcal{E}_1$.

Proposition 3.1: If $\mathcal{F}$ is a functional in $\mathcal{C}(H_1)$, then $\mathcal{F} \circ B$ is a functional in $\mathcal{C}(H)$. Furthermore, if $\mathcal{F} \in \mathcal{L}^1(H_1)$, then $\mathcal{F} \circ B \in \mathcal{L}^1(H)$ if and only if B^*B is of the form "$\mathbb{1}$ + trace class operator". The following relation then holds:

$$\int_{H_1} \mathcal{F}(h_1)d[h_1] = \text{Det}(B^*B)^{\frac{1}{2}} \int_H \mathcal{F}(Bh)d[h]. \tag{3.2}$$

Proof: The proof goes as in Lemma 3.2, using the fact that

$$B \circ \phi_i(x) = B(xe_i) = \sqrt{\lambda_i}xe_i^1 = \phi_i^1(\sqrt{\lambda_i}x)$$

where ϕ_I^1 is the map constructed as in (3.1) from $\mathbb{R}^{|I|}$ into H_1. ∎

In the context of strings, the operators under which the formal Lebesgue measures are transformed are in general built up from elliptic operators which require a heat-kernel regularization.

Let E and E_1 be two vector bundles over a smooth compact boundaryless surface S, equipped with scalar products $< \cdot, \cdot >$ and $< \cdot, \cdot >_1$. Let H and H_1 be the closures of the spaces of smooth sections of these bundles, which we shall assume to be separable. Let now B be an injective elliptic operator from E onto E_1. The operator B being densely defined has a well-defined adjoint B^*, and B^*B is a positive self-adjoint elliptic operator on a compact surface. It has a purely discrete spectrum $(\lambda_n)_{n \in \mathbb{N}}$ and we shall choose the orthonormal basis $\mathcal{E}$ of H to be an orthonormal basis of eigenvectors of B^*B. As above, we build the corresponding orthonormal basis $\mathcal{E}_1$ of H_1 and define the sets $\mathcal{C}(H_1)$ and $\mathcal{L}^1(H_1)$. Using the results of section I.5, for $\varepsilon > 0$, we can define the operator $h_\varepsilon(B^*B)$ which is of the form "$\mathbb{1}$ + trace class".

Since the operator B^*B is not of the form "$\mathbb{1}$ + trace class", we cannot strictly relate formal Lebesgue integrals on H_1 to formal Lebesgue integrals on H as in (3.2), but we can define cutoff Lebesgue integrals on H_1 as follows:

Definition 3.3: Let $\mathcal{F}$ be a functional in $\mathcal{C}(H_1)$ such that $\mathcal{F} \circ B \in \mathcal{L}^1(H)$. We define a "regularized" Lebesgue integral of $\mathcal{F}$ on H_1 by

$$\int_{E_1} \mathcal{F}(H_1)d_{\text{reg}}[h_1] \equiv \text{Det}(B^*B)^{\frac{1}{2}} \int_E \mathcal{F}(Bh)d[h] \tag{3.3}$$

where Det is to be understood in the sense of section I.5.

Formula (3.3) is relevant for the Faddeev–Popov procedure in gauge theories. If $P \to P/G$ describes a principal fiber bundle with gauge group G as in section I.6, then a reparametrization of P in terms of a section S and the gauge group G gives rise to the Faddeev–Popov operator F_p from $E_1 \equiv$

II.3: Formal Lebesgue measures on Hilbert spaces

$T_eG \times W_p$ to $E = T_pP$ with the notation of section I.6 . The functional quantization of the model involves functional integrals of the form studied above, namely $\int_{E_1} \mathcal{F}(h_1)d[h_1]$, where $\mathcal{F}$ is a functional on H_1. Assuming that the Faddeev–Popov operator is elliptic (in fact the situation is more complicated, as we saw in section I.6, since the Faddeev–Popov operator is built up from elliptic operators but is itself not elliptic; formally, one extends this description for elliptic operators to the Faddeev–Popov operators) we can apply definition 3.3 and define

$$\int_{E_1} \mathcal{F}(h_1)d_{\text{reg}}[h_1] = \text{Det}(F_p^*F_p)^{\frac{1}{2}} \int_E \mathcal{F}(F_ph)d[h] \qquad (3.4)$$

whenever $\mathcal{F} \circ F_p$ lies in $\mathcal{L}^1(H)$. We recall that Det $(F_p^*F_p)$ has been defined in Def. I.6.1 as a limit when $\varepsilon \to 0$ involving Det $(F_p^*F_p)_\varepsilon$ with the divergences compensated by a counterterm $N_\varepsilon(p)$. This is the renormalization procedure. In order to analyse this compensation of divergences, we shall work with ε-cutoff measures instead of the previous "regularized" measures (i.e we do not take the limit of the cut-off determinants), namely we define for $\varepsilon > 0$ an ε-cutoff transformed measure

$$\int_{E_1} \mathcal{F}(h_1)d_\varepsilon[h_1] = \text{Det}_\varepsilon(F_p^*F_p)^{\frac{1}{2}} \int_E \mathcal{F}(F_ph)d[h] \qquad (3.5)$$

so that we can write

$$\int_{E_1} \mathcal{F}(h_1)d_{\text{ren}}[h_1] = \lim_{\varepsilon \to 0} N_\varepsilon(p) \int_{E_1} \mathcal{F}(h_1)d_\varepsilon[h_1]$$

where $N_\varepsilon(p)$ is the compensating term chosen such that

$$\lim_{\varepsilon \to 0} \left(\exp[\log \text{Det}_\varepsilon(F_p^*F_p) - N_\varepsilon(p)] \right)^{\frac{1}{2}} = \text{Det} F_p. \qquad (3.6)$$

The renormalization term will be an ε-dependent term, the inverse of which diverges when ε tends to zero, in the same way as the heat-kernel determinants do. In the case of strings, setting $p = g$, $N_\varepsilon(p) = \mu_\varepsilon^2 \int_S \sqrt{\det g}dz$, and $e^{-\mu_\varepsilon^2 \int_S \sqrt{\det g}dz}$ is called the cosmological term.

II.4 The Gaussian integration on the space of embeddings

The first step of the formal functional quantization for bosonic strings was the computation of a Gaussian integral on the space of smooth embeddings [†] giving rise to an infinite-dimensional determinant (see II.2.6), namely

$$Z_g \equiv \int_{\mathcal{F}} e^{-D(X,g)} d_g[X] = \left[\frac{\text{``det'''}(-\Delta_g)}{\int_S \sqrt{\det g}\, dz}\right]^{-\frac{d}{2}} \quad (4.1)$$

where $D(X, g)$ is the classical action, $d_g[X]$ is a formal Lebesgue measure on the space of embeddings [†], and "det'" is a formal infinite dimensional determinant. In this section, we shall give a description of this formal Gaussian integration using the rules for formal Lebesgue integration on Hilbert spaces described in section II.3.

We first recall the expression of the classical action for bosonic strings. Here S denotes as before a smooth compact boundaryless surface. For a smooth metric $g \in \mathcal{M}$ (see (I.2.6)), the classical action for bosonic strings is given by (I.1.4), namely

$$D(X, g) = \frac{1}{2} \int_S \sqrt{\det g}\, g^{ij} \frac{\partial X^\alpha}{\partial z^i} \frac{\partial X^\alpha}{\partial z^j} dz \quad (4.2)$$

where $X = (X^\alpha)_{\alpha=1,\ldots,d}$ is a Sobolev map of S into $\mathbb{R}^d$.

Let Δ_g (see (I.4.7)) denote the Laplace–Beltrami operator on $C^\infty(S, \mathbb{R})$ and let $(\cdot, \cdot)_g$ denote the scalar product induced by the metric g on $C^\infty(S, \mathbb{R})$ (see (I.2.3)). Then the action reads

$$D(X, g) = -\frac{1}{2} \sum_{\alpha=1}^d (\Delta_g X^\alpha, X^\alpha)_g. \quad (4.3)$$

Since Δ_g is an elliptic operator on a compact surface, it leaves the space $(\ker \Delta_g)^\perp$ orthogonal to its kernel invariant and we can define

$$\Delta'_g \equiv \Delta_g|_{(\ker \Delta_g)^\perp}.$$

Accordingly we split $X^\alpha \in C^\infty(S, \mathbb{R})$ into an orthogonal sum:

$$X^\alpha = X^{\alpha\prime} + X_0^\alpha \quad \text{with} \quad \begin{aligned} X^{\alpha\prime} &\in (\ker \Delta_g)^\perp \\ X_0^\alpha &\in \ker \Delta_g \cong \mathbb{R}. \end{aligned}$$

[†] The integration should actually take place over the larger space of Sobolev mappings of S into $\mathbb{R}^d$, and these mappings need not necessarily be embeddings.

II.4: The Gaussian integration on the space of embeddings   107

The action (4.3) reads

$$D(X,g) = -\frac{1}{2}\sum_{\alpha=1}^{d}(\Delta'_g X^{\alpha'}, X^{\alpha'})_g. \tag{4.4}$$

Let H' be the closure of $C^\infty(S,\mathbb{R})$ w.r.t. the scalar product (I.2.3). Since the operator Δ'_g is elliptic on a compact boundaryless surface S, it has a pure point spectrum and we shall take $\mathcal{E} = (e_n)_{n\in\mathbb{N}}$ to be an orthonormal basis of eigenvectors of Δ'_g. Following section II.3, we can then define integrals of the form

$$\int_{H'} F'(h)d_g[h] = \lim_{I\to\mathbb{N}}\left(\frac{1}{2\pi}\right)^{\frac{|I|}{2}}\int \mathcal{F}'\circ\phi_I(h_{i_1},\ldots,h_{i_{|I|}})dh_{i_1}\ldots dh_{i_{|I|}} \tag{4.5}$$

where ϕ_I is the map defined in (II.2.1) w.r.t. the basis $\mathcal{E}$ where $\mathcal{F}'$ is in $\mathcal{L}^1(H')$. The formal Lebesgue measure $d_g[X]$ on H' is indexed by g since the underlying scalar product $(\cdot,\cdot)_g$ is g-dependent.

On leaving out the integration on the kernel of Δ_g, the formal expectation on the l.h.s. of (4.1) corresponds to the following choice of the functional $\mathcal{F}'$:

$$\mathcal{F}'(X) = e^{\frac{1}{2}\sum_{\alpha=1}^{d}(\Delta'_g X^\alpha, X^\alpha)}. \tag{4.6}$$

Now taking into account the integration on $\ker\Delta_g$, the partition function Z_g can heuristically be written

$$Z_g = \prod_{\alpha=1}^{d}\left[\int_{(\ker\Delta_g)^\perp}e^{\frac{1}{2}(\Delta'_g X^{\alpha'}, X^{\alpha'})_g}d_g[X'^\alpha]\int_{\ker\Delta_g}d_g[X_0^\alpha]\right]$$

where $d_g[X_0^\alpha]$ is the Lebesgue measure on the 1-dimensional vector space $\ker\Delta_g$. Since $\ker\Delta_g$ is 1-dimensional (isomorphic to $\mathbb{R}$), the second integral is a volume of a vector space and hence ill defined. We can overcome this difficulty by introducing a regularized version of the partition function Z_g as follows.

Since $-\Delta_g$ is a positive self-adjoint elliptic operator on a compact boundaryless surface S, for $\varepsilon > 0$ we can define the operator $h_\varepsilon(-\Delta'_g)$ where h_ε is given by (I.5.16). For $\varepsilon > 0$, $X \in C^\infty(S,\mathbb{R}^d)$, $g \in M$, let us define the ε-cutoff action

$$D_\varepsilon(X,g) := \frac{1}{2}\sum_{\alpha=1}^{d}(h_\varepsilon(-\Delta'_g)X'^\alpha, X'^\alpha)_g \tag{4.7}$$

and the corresponding ε-cutoff partition function:

$$Z_g^\varepsilon := \left(\prod_{\alpha=1}^d \int_{(\ker \Delta_g)^\perp} e^{-\frac{1}{2}(h_\varepsilon(-\Delta_g')X^{\alpha'},X^{\alpha'})_g} d_g[X'^\alpha]\right) \left(\sqrt{\mathrm{Vol}_g(S)}\right)^d. \quad (4.8)$$

Set

$$\mathcal{F}_\varepsilon^\alpha(X'^\alpha) := e^{-\frac{1}{2}(h_\varepsilon(-\Delta_g')X'^\alpha,X'^\alpha)_g}$$

for $X'^\alpha \in (\ker \Delta_g)^\perp$. Since the operator $h_\varepsilon(-\Delta_g')$ is of the form "$\mathbb{1}$ + trace class", by Lemma 3.2, $\mathcal{F}_\varepsilon^\alpha \in L^1(H')$ and we have

$$\int_{H'} \mathcal{F}_\varepsilon^\alpha(X'^\alpha) d_g[X'^\alpha] = \mathrm{Det}(h_\varepsilon(-\Delta_g'))^{-\frac{1}{2}} \quad (4.9)$$

(where "Det" was defined in section I.5). This leads to the final expression of the ε-cutoff partition function:

$$Z_g^\varepsilon = \left(\frac{\mathrm{Det}(h_\varepsilon(-\Delta_g'))}{\mathrm{Vol}_g(S)}\right)^{-\frac{d}{2}}. \quad (4.10)$$

We have therefore replaced the formal ill-defined Gaussian integration (4.1) by the well defined expression

$$Z_g^\varepsilon = \left(\frac{\mathrm{Det}_\varepsilon'(-\Delta_g)}{\int_S \sqrt{\det g}\, dz}\right)^{-\frac{d}{2}} \quad (4.11)$$

where $\mathrm{Det}_\varepsilon'$ was defined in section I.5. Heuristically

$$Z_g^\varepsilon = (\mathrm{Vol}_g(S))^{\frac{d}{2}} \int_{H'} e^{-D_\varepsilon(X,g)} d_g[X'].$$

II.5 The Faddeev–Popov procedure for bosonic strings

As we saw in section II.4, after integration on the space of embeddings, the ε-cutoff Polyakov measure $d\nu_\varepsilon[g]$ reduces to a formal measure on the manifold $\mathcal{M}$ of smooth metrics:

$$d\nu_\varepsilon[g] = \frac{1}{Z_\varepsilon}\mathcal{F}_\varepsilon(g)d[g] \qquad (5.1)$$

where

$$Z_\varepsilon \equiv \int_{\mathcal{M}} \mathcal{F}_\varepsilon(g)d[g]$$

and

$$\mathcal{F}_\varepsilon(g) = \left(\frac{\mathrm{Det}'_\varepsilon(-\Delta_g)}{\int_S \sqrt{\det g}\,dz}\right)^{-\frac{d}{2}} e^{-\mu_\varepsilon^2 \int_\Lambda \sqrt{\det g(z)}\,dz}$$

Here $\mathcal{M}$ denotes as before (see (I.2.6)) the manifold of smooth Riemannian metrics on a smooth compact boundaryless surface S of genus $p > 1$.

According to the results of section I.3, the manifold $\mathcal{M}$ is C^∞ diffeomorphic to a product of spaces

$$\begin{aligned} T_p \times \mathcal{D}_0 \times C &\to \mathcal{M} \\ (t, f, \varphi) &\to f^* e^\varphi g_t \end{aligned} \qquad (5.2)$$

where T_p is the Teichmüller space for the surface S, $\mathcal{D}_0$ is the group of smooth diffeomorphisms homotopic to the identity and C is the conformal group as in section I.3. The map

$$\begin{aligned} \sigma_h : T_p &\to M_{-1} \\ t &\to g_t \end{aligned}$$

is the global harmonic gauge defined in (I.3.19). We shall set $S \equiv \sigma_h(T_p) = \{g_t, t \in T_p\}$. According to the formal transformation rule (II.3.5), the measure (5.1) then transforms into an integral on $T_p \times \mathcal{D}_0 \times C$:

$$d\nu_\varepsilon[t, f, \varphi] = \frac{1}{Z_\varepsilon}\mathcal{F}_\varepsilon(t, f, \varphi)\mathrm{Det}_\varepsilon(F_g^* F_g)^{\frac{1}{2}} dt\, d_g[f]\, d_g[\varphi] \qquad (5.3)$$

giving rise to the ε-cutoff Faddeev–Popov determinant $\mathrm{Det}_\varepsilon(F_g)$ which was described in section I.6, taking $p = g$. Here dt is the volume measure on T_p, $d_g[f]$ (resp. $d_g[\varphi]$) are formal volume measures on $\mathcal{D}_0$ resp. C, arising from the Riemannian structure defined in terms of the metric g, namely the one which induces the scalar product (I.2.26) (resp. (1.2.3)) on the tangent space. When there is no ambiguity about the choice of the underlying metric "g", we shall sometimes omit the index "g" in the subsequent chapters.

II.5: The Faddeev-Popov procedure for bosonic strings

Our aim in this section is to describe this transformation more precisely and to compute the Faddeev–Popov determinant in the case of bosonic strings (see also [K], [Jas1], [Jas2] for related works). For this, we shall essentially use the framework and results of sections I.6 and II.3.

From the results of section I.3, $\mathcal{M}$ is a smooth trivial principal fibre bundle over T_p with structure group $\mathcal{D}_0 \odot C$ (see Theorem I.3.5 and the proof of this theorem for the global triviality).

We set $P \equiv \mathcal{M}$, $G \equiv \mathcal{D}_0 \odot C$ with the notation of section I.6. Here, the semiproduct law is given by

$$(f_1, e^{\varphi_1}) \cdot (f_2, e^{\varphi_2}) = (f_1 \circ f_2, (e^{\varphi_1} \circ f_2)e^{\varphi_2}).$$

Notice that Hyp. 1 of section I.6 is fulfilled since the quotient space $P/G = \mathcal{M}/\mathcal{D}_0 \odot C \simeq T_p$ has dimension $6p - 6$ over $\mathbb{R}$. Indeed, by Theorem I.3.2, we have $\dim T_p = \dim Q(\Sigma)$, and by (I.7.43) and (I.7.40) we have

$$\dim_{\mathbb{C}} Q(\Sigma) = \dim_{\mathbb{C}} H^0(\Sigma', \Omega^2)^* = \dim_{\mathbb{R}} H^0(\Sigma, \Omega^2)^* = 3(p-1)$$

Moreover Hyp. 2 is fulfilled since $G = \mathcal{D}_0 \odot C$ is an infinite-dimensional Fréchet manifold with tangent space $C^\infty(TS) \times C^\infty(S, \mathbb{R})$, so that $E = TS \times (S \times \mathbb{R})$.

The group G acts smoothly on $\mathcal{M}$ by the right action

$$\begin{aligned}\mathcal{D}_0 \times C \times \mathcal{M} &\to \mathcal{M} \\ (f, e^\varphi, g) &\to f^* e^\varphi g.\end{aligned} \qquad (5.4)$$

Since the infinite-dimensional Fréchet manifold $\mathcal{M}$ is modeled on the space of smooth sections $C^\infty(S^2 T^*)$ of the bundle of symmetric 2 covariant tensors $S^2 T^*$, the tangent map to the map

$$\begin{aligned}\theta_g : \mathcal{D}_0 \odot C &\to \mathcal{M} e_g^\varphi \\ [f, e^\varphi g] &\to f^*\end{aligned}$$

at the point (f, e^φ) is given for a metric g by $D(f, e^\varphi)\theta_g = f^* \tau_{\bar{g}}$ where $\bar{g} \equiv e^\varphi g$ and

$$\begin{aligned}\tau_{\bar{g}} : C^\infty(TS) \times C^\infty(S, \mathbb{R}) &\to C^\infty(S^2 T^*) \simeq T_{\bar{g}} \mathcal{M} \\ (u, \lambda) &\to \nabla_{\bar{g}} u + \lambda \bar{g}\end{aligned}$$

where $\nabla_{\bar{g}}$ denotes the Lie derivative w.r.t. the metric g, $(\nabla_{\bar{g}} u)_{ab} = \nabla_a u_b + \nabla_b u_a$.

We are therefore in the case studied in section I.6.3, where

$$H = \mathcal{D}_0, \ K = C.$$

II.5: The Faddeev-Popov procedure for bosonic strings

When equipped with the scalar product given by (I.2.11), the space $T_{\bar{g}}\mathcal{M} \simeq C^{\infty}(S^2T^*)$ splits into two orthogonal spaces, namely the space of pure trace tensors $C^{\infty}(S, \mathbb{R}) \cdot g$ and the space of tracefree tensors $C^{\infty}(S_0^2T^*)$. The image of the space $C^{\infty}(S, \mathbb{R})$ under $\tau_{\bar{g}}$ coincides with the space of pure trace tensors and the operator $\tau_{\bar{g}}$ can, as in (I.6.13), be seen as a matrix operator acting from $C^{\infty}(T\Lambda) \times C^{\infty}(S, \mathbb{R})$ into $C^{\infty}(S_0^2T^*) \times C^{\infty}(S, \mathbb{R}) \cdot \bar{g}$, so that we can write

$$\tau_{\bar{g}} = \begin{bmatrix} A_{\bar{g}} & 0 \\ C_{\bar{g}} & B_{\bar{g}} \end{bmatrix}. \tag{5.5}$$

The operators $A_{\bar{g}}$, $B_{\bar{g}}$ and $C_{\bar{g}}$ are defined as in section I.6; let $\pi_{\bar{g}}^K$ denote the orthogonal projection onto the tracefree tensors. Then

$$A_{\bar{g}} = \pi_{\bar{g}}^K \nabla_{\bar{g}} = P_{\bar{g}}; \quad C_{\bar{g}} = \frac{1}{2}\text{tr}_{\bar{g}}(\nabla_{\bar{g}}(\cdot)) \cdot \bar{g}; \quad B_{\bar{g}}\lambda = \lambda \cdot \bar{g}$$

where for a metric g, $P_{\bar{g}}$ was defined in (I.4.9). The operator $B_{\bar{g}}$ is an isometry from $C^{\infty}(S, \mathbb{R})$ equipped with the scalar product (I.2.3) into $T_{\bar{g}}\mathcal{M}$ equipped with the scalar product (I.2.11) and, as in section I.6, we shall identify the operator $\tau_{\bar{g}}$ with the matrix operator

$$\tau_{\bar{g}} = \begin{bmatrix} P_{\bar{g}} & 0 \\ \frac{1}{2}\text{tr}_{\bar{g}}(\nabla_{\bar{g}}(\cdot)) \cdot \bar{g} & \mathbb{1} \end{bmatrix}. \tag{5.6}$$

Since the operator $P_{\bar{g}}$ is injective for genus $p > 1$ (see I.7.42), so is the operator $\tau_{\bar{g}}$. Moreover, the operator $P_{\bar{g}}$ being a differential operator with an injective symbol, so is the matrix operator $\tau_{\bar{g}}$ (see [Eb]).

We can now apply the results of section I.6.

For $g = f^*e^{\varphi}g_t \in \mathcal{M}$, the Faddeev–Popov operator is given by (I.6.8) on setting $p = g$, $a = (f, e^{\varphi})$, $x = g_t \in \Sigma$. Now, applying the results of Proposition I.6.2, and taking the point $p \in P$ to be the metric $g = f^*e^{\varphi}g_t \in \mathcal{M}$, and $\pi_{\bar{g}}$ to be the orthogonal projection onto the image of $\tau_{\bar{g}_t}$, we find that the ε-cutoff Faddeev-Popov determinant for strings reads (see Def. I.6.2 and Prop. I.6.2):

$$\text{Det}_{\varepsilon}(F_g^*F_g)^{\frac{1}{2}} = \text{Det}'_{\varepsilon}(-\Delta_{\bar{g}_t})\text{Det}_{\varepsilon}(P_{\bar{g}_t}^*P_{\bar{g}_t})^{\frac{1}{2}} \frac{\det(<\Psi_{\bar{g}_t}^i, \xi_{\bar{g}_t}^j>_{\bar{g}_t})}{\det(<\Psi_{\bar{g}_t}^i, \Psi_{\bar{g}_t}^j>_{\bar{g}_t})^{\frac{1}{2}}}. \tag{5.7}$$

Thus the formal ε-cutoff Polyakov measure (5.1) reads for $g = f^*e^{\varphi}g_t$:

$$d\nu_{\varepsilon}[t, f, \varphi] = \frac{1}{Z_{\varepsilon}} e^{-\mu_{\varepsilon}^2 \int_{\Lambda} \sqrt{\det \bar{g}_t(z)} dz} \left(\frac{\text{Det}'_{\varepsilon}(-\Delta_{\bar{g}_t})}{\int_{\Lambda} \sqrt{\det \bar{g}_t(z)} dz} \right)^{\frac{-d}{2}}$$

$$\text{Det}_{\varepsilon}(P_{\bar{g}_t}^*P_{\bar{g}_t})^{\frac{1}{2}} \frac{\det(<\Psi_{\bar{g}_t}^i, \xi_{\bar{g}_t}^j>_{\bar{g}_t})}{\det(<\Psi_{\bar{g}_t}^i, \Psi_{\bar{g}_t}^j>_{\bar{g}_t})^{\frac{1}{2}}} dt d_{\bar{g}_t}[f] d_{\bar{g}_t}[\varphi]. \tag{5.8}$$

II.5: The Faddeev-Popov procedure for bosonic strings

Notice that we have used the fact that all the quantities involved in the integrand are invariant under the action of the diffeomorphism group since the underlying scalar products are invariant and so are the determinants as products of eigenvalues. On formally integrating with respect to $d_{\bar{g}_t}[f]$ along the diffeomorphisms, the new formal ε-cutoff Polyakov measure on $T_p \times C$ reads

$$d\mu_\varepsilon[\varphi, t] \equiv \frac{1}{Z_\varepsilon} e^{-\mu_\varepsilon^2 \int_S \sqrt{\det \bar{g}_t}\, dz} \left(\frac{\text{Det}'_\varepsilon(-\Delta_{\bar{g}_t})}{\int_S \sqrt{\det \bar{g}_t}\, dz} \right)^{\frac{-d}{2}}$$

$$\text{Det}_\varepsilon(P^*_{\bar{g}_t} P_{\bar{g}_t})^{\frac{1}{2}} \frac{\det(<\Psi^i_{\bar{g}_t}, \xi^j_{\bar{g}_t}>_{\bar{g}_t})}{(\det <\Psi^i_{\bar{g}_t}, \Psi^j_{\bar{g}_t}>_{\bar{g}_t})^{\frac{1}{2}}}\, dt\, d_{\bar{g}_t}[\varphi], \qquad (5.9)$$

where Z_ε is the corresponding partition function defined as usual, i.e. Z_ε is the normalization constant heuristically making $D\mu_\varepsilon[\varphi, t]$ into a probability measure.

II.6 The Polyakov measure in noncritical dimension and the Liouville measure

In section II.5, we defined the ε-cutoff Polyakov measure (see (II.5.9))

$$d\mu_\varepsilon[\varphi,t] = \frac{1}{Z_\varepsilon} e^{-\mu_\varepsilon^2 \int_S \sqrt{\det \bar{g}_t} dz} \left[\frac{\mathrm{Det}'_\varepsilon(-\Delta_{\bar{g}_t})}{\int_S \sqrt{\det \bar{g}_t} dz}\right]^{\frac{-d}{2}}.$$

$$\mathrm{Det}_\varepsilon(P^*_{\bar{g}_t} P_{\bar{g}_t})^{\frac{1}{2}} \frac{\det <\Psi^i_{\bar{g}_t}, \xi^j_{\bar{g}_t}>_{\bar{g}_t}}{(\det <\Psi^i_{\bar{g}_t}, \Psi^j_{\bar{g}_t}>_{\bar{g}_t})^{\frac{1}{2}}} dt d_{\bar{g}_t}[\varphi]. \qquad (6.1)$$

The aim of this section is to give an interpretation of the integration along the conformal factor φ in terms of a family of well-defined "Liouville measures" on curved space as described in section I.10, parametrized by t, which we shall formally denote by $e^{-\mathcal{L}(\varphi,t)}d[\varphi]$ for the moment. We show that when $\varepsilon \to 0$, the regularized Polyakov measure reads

$$d\mu[\varphi,t] = \frac{1}{Z} e^{-\mathcal{L}(\varphi,t)} h(t) d[\varphi] dt,$$

$$\text{where} \quad Z = \int e^{-\mathcal{L}(\varphi,t)} h(t) d[\varphi] dt. \qquad (6.2)$$

In this section, we shall therefore concentrate on the φ integration and leave aside for the moment the integration on Teichmüller space which will be more closely described in the next section. Let us extract the φ-dependent terms from the formal density of $d\mu[\varphi,t]$ with respect to $D[\varphi]dt$.

For this, we first notice that the expression $\det(<\Psi^i_{\bar{g}_t}, \xi^j_{\bar{g}_t}>_{\bar{g}_t})$ is independent of the conformal factor φ. We recall that $\ker P^*_g = (\ker \partial_2)^*$ (see (I.7.43)) depends only on the conformal class of g, i.e. $\ker P^*_g = \ker P^*_{e^\varphi g}$ for $\varphi \in C^\infty(S, \mathbb{R})$. The basis $(\Psi^i_g)_{i=1,\ldots,6p-6}$ of $\ker P^*_g$ can therefore be chosen independently of the conformal factor φ arising in the description of $\mathcal{M}$ by $g = f^* e^\varphi g_t$ given in (II.5.2), so that finally $(\Psi^i_{\bar{g}_t}) = (\Psi^i_{g_t})$.

On the other hand, the basis $\xi^i_g = \partial_{t_i} g$ of W_g (see (I.6.2) with p=g) where $t_i, i = 1, \ldots, 6p-6$, clearly has the property $\xi^i_{e^\varphi g} = e^\varphi \xi^i_g$ so that $\xi^i_{\bar{g}_t} = e^\varphi \xi^i_{g_t}$. But since $<h,k>_{e^\varphi g} = <e^{-\varphi}h,k>_g$ for covariant 2-tensors (see formula (I.2.11)), we finally have that

$$\det(<\Psi^i_{\bar{g}_t}, \xi^j_{\bar{g}_t}>_{\bar{g}_t}) = \det(<\Psi^i_{g_t}, \xi^j_{g_t}>_{g_t})$$

is independent of φ.

Since in this section we are concerned with the φ-dependence of the formal density of μ_ε, we shall concentrate on the terms in the integrand in (6.1) which do not involve Ψ and effectively depend on φ.

From the results on the asymptotics of the heat-kernel expansion for elliptic operators described in section I.11 (see in particular (I.11.19)), it follows that

$$d\mu_\varepsilon[\varphi, t] = \frac{1}{Z_\varepsilon} e^{-\mu_\varepsilon^2 \int_S \sqrt{\det \tilde{g}_t}\, dz} \exp(F_d(\varepsilon, \varphi)) h_\varepsilon(t) dt d[\varphi] \qquad (6.3)$$

where $h_\varepsilon(t)$ is an expression which depends only on t and ε and is defined by

$$h_\varepsilon(t) = \det <\Psi^i_{g_t}, \Psi^j_{g_t}>_{g_t}^{-\frac{1}{2}} \left(\frac{\operatorname{Det}'_\varepsilon(-\Delta_{g_t})}{\int_S \sqrt{\det g_t}\, dz}\right)^{\frac{-d}{2}} \operatorname{Det}_\varepsilon(P^*_{g_t} P_{g_t})^{\frac{1}{2}}$$
$$\cdot \det <\Psi^i_{g_t}, \xi^j_{g_t}>_{g_t} . \qquad (6.4)$$

Here, as in section I.11,

$$F_d(\varepsilon, \varphi) = \frac{d-2}{8\pi\varepsilon} \left[\int_S \sqrt{\det g_t}\, e^\varphi dz - \int_S \sqrt{\det g_t}\, dz\right]$$
$$+ \frac{d-26}{48\pi} \left(\frac{1}{2} \int_S \sqrt{\det g_t}\, \varphi(-\Delta_{g_t}\varphi) dz + \int_S \sqrt{\det g_t}\, \varphi R_{g_t}\, dz\right) \qquad (6.5)$$
$$+ K(\varepsilon, \varphi)$$

where $K(\varepsilon, \varphi) \to 0$ as $\varepsilon \to 0$ and where R_g is the scalar curvature of the metric g.

Let us now choose the cosmological constant μ_ε in (6.3) so as to cancel the terms diverging as $\varepsilon \downarrow 0$ occuring in (6.4):

$$\mu_\varepsilon^2 \equiv \frac{d-2}{8\pi\varepsilon} - \frac{d-26}{48\pi}\lambda \qquad (6.6)$$

where λ is a positive constant. Then, letting ε go to zero, the renormalized measure $d\mu_{\text{ren}}[\varphi, t]$ (heuristic limit of (6.3) as $\varepsilon \downarrow 0$) reads:

$$d\mu_{\text{ren}}[\varphi, t]$$
$$= \frac{1}{Z_{\text{ren}}} e^{\frac{d-26}{48\pi}\left[\frac{1}{2}\int_S \sqrt{\det g_t}\, \varphi(-\Delta_{g_t}\varphi) dz + \int_S \sqrt{\det g_t}\, \varphi R_{g_t}\, dz + \lambda \int_S \sqrt{\det g_t}\, e^\varphi dz\right]} \qquad (6.7)$$
$$h(t) dt d[\varphi]$$

where

$$h(t) = \lim_{\varepsilon \to 0} h_\varepsilon(t) \cdot e^{\frac{2-d}{8\pi\varepsilon} \int_S \sqrt{\det g_t}\, dz} \qquad (6.8)$$

is a finite expression (see section I.5).

II.6: The Polyakov measure in noncritical dimension

Let us now set
$$\alpha \equiv \sqrt{\frac{48\pi}{26-d}}$$
and make the formal change of variable:
$$\psi \equiv \alpha^{-1}\varphi$$
in (6.7). Then since the metrics g_t have curvature -1, the formal renormalized measure $d\mu_{\text{ren}}$ reads:

$$d\mu_{\text{ren}}[\psi, t] = \frac{1}{Z_{\text{ren}}} e^{-\mathcal{L}(\psi,t)} h(t) d[\psi] dt, \tag{6.9}$$

whereby

$$\mathcal{L}(\psi, t) \equiv \frac{1}{2} \int_S \psi(-\Delta_{g_t}\psi)\sqrt{\det g_t} dz + \lambda \int_S \sqrt{\det g_t} e^{\alpha\psi} dz - \frac{1}{\alpha} \int_S \sqrt{\det g_t} \psi dz, \tag{6.10}$$

which is the Lagrangian of the Liouville model (described in section I.10) on the surface S indexed by a Teichmüller parameter corresponding to the massless free bosonic field in two dimensions perturbed by an exponential interaction and a linear interaction.

Liouville measures on curved space were given a precise mathematical description in section I.10, formula (I.10.28), the potential being given by (I.10.27). With the notation of section I.10, whenever

$$\alpha^2 < 4\pi, \quad \text{i.e.} \quad d \leq 13,$$

we can interpret the $e^{-\mathcal{L}(\psi,t)} d[\psi]$ measure in (6.9) as the probability measure given by the Liouville measure $\nu^c_{\alpha,t}$ built as in (I.10.28), the underlying metric g being here replaced by g_t (which makes the measure t-dependent). Thus we have the following interpretation of the formal renormalized Polyakov measure :

$$d\mu_{\text{ren}}[g_t] \equiv \frac{1}{Z_{\text{ren}}} \left(\frac{\text{Det}'(-\Delta_g)}{\int_S \sqrt{\det g_t} dz} \right)^{-\frac{d}{2}} \text{Det}(P_{g_t}^* P_{g_t})^{\frac{1}{2}} (\det(\Psi^i_{g_t}, \Psi^j_{g_t}))^{-\frac{1}{2}} dt d[\psi]$$

$$= \left(\int_{T_p} h(t) dt \right)^{-1} d\nu^c_{\alpha,t}(\psi) h(t) dt. \tag{6.11}$$

The t-integration $h(t)dt$ is here still a heuristic one but will be given a precise meaning in the next section.

Remarks:

(1) It is important to notice at this point that if $d = 26$, that is to say if the dimension is <u>critical</u>, the φ-dependent term in (6.7) vanishes, so that there are no φ-dependent factors in the formal Polyakov measure at critical dimension and the measure formally reduces to a measure on Teichmüller space.

(2) Introducing the cosmological term $e^{-\mu_\epsilon^2 \int_S \sqrt{\det g} dz}$ boils down to taking the renormalized limit of the heat-kernel determinants involved. We could also have chosen to take directly the zeta-function regularized determinants. These differ from the heat-kernel regularized ones by a constant (depending only on the topology of S) (see (I.5.22)) and this choice therefore has no influence on the conformal factors.

(3) In [AHKPS2] there is a discussion of a renormalization term of the form $\int_S e^{\mu_\epsilon^2 \sqrt{g} dz}$ where μ_ϵ is a function on S defined by

$$\mu_\epsilon^2(\eta) = \frac{d-2}{8\pi\epsilon} - \lambda \frac{d-26}{8\pi\epsilon} \exp\left[-\frac{1}{2}\frac{48\pi}{26-d} < \chi_\epsilon^\eta, G_{g_t}\chi_\epsilon^\eta >_{g_t}\right],$$

with $\chi_\epsilon^\eta = e^{\epsilon \Delta_{g_t}}(\cdot, \eta)$,

$$G_{g_t} = (\Delta'_{g_t})^{-1}, \quad \lambda > 0.$$

The conclusions remain true also with this renormalization. See [AHKPS2] for a more detailed discussion concerning this point.

(4) If one takes into account the φ-dependence of the measure $d_{\bar{g}_t}[\varphi]$ (recall that $\bar{g}_t = e^\varphi_{g_t}$), then (6.6) becomes

$$\mu_\epsilon^2 = \frac{d-2}{8\pi\epsilon} - \frac{d-25}{48\pi}\lambda.$$

This point is discussed in [AHKPS3].

II.7 The Polyakov measure in the critical dimension $d = 26$

We recall that in section II.4 we replaced the integration over the embeddings X by the determinant of the Laplace operator, and that in section II.5 we implemented the Faddeev–Popov procedure to discard the integration over D_0 (the group of smooth diffeomorphisms homotopic to the identity) by introducing the Faddeev–Popov determinant, which essentially coincides with the determinant of the operator P_g given by (I.4.10). In section II.6, we interpreted the integration over the conformal deformations in terms of a family of Liouville measures indexed by Teichmüller parameters. In this section we shall investigate the integration over Teichmüller space; the measure on the latter is given by the Weil–Petersson metric. There are many articles concerning a "direct" description of the Polyakov measure in critical dimension, e.g. [BJ], [BK], [BM], but these do not discuss, as we did, its derivation from functional integrals considered for all d.

As before, we let $T_{p,k}$ be the Teichmüller space of surfaces of genus p with k boundaries. As discussed in section I.7, it can be considered as a totally real subspace of T_q, with $q = 2p + k - 1$, the genus of the Schottky double. We recall that the cotangent space of $T_{p,k}$ is given by the holomorphic quadratic differentials which are real on the boundary, and that on this space we have the Weil–Petersson metric. Complexifying, we obtain the space of holomorphic quadratic differentials on closed surfaces of genus q, and these differentials vary holomorphically with the conformal structure of the surface. We let $q_1, ..., q_{q-3}$ be a basis of the space of holomorphic quadratic differentials, real on the boundary, for elements of $T_{p,k}$. Over $\mathbb{C}$, they yield a basis for elements of T_q. Recall that the kernel of the operator P_g^* can be identified with the dual of the space of quadratic differentials (see (I.7.43)) and hence the term $\det < \Psi_g^i, \Psi_g^j >_g$ arising in the expression of the Polyakov measure after the Faddeev–Popov procedure has been implemented coincides with the determinant $\det < q_i, q_j >_g$.

Throughout this section, we shall use the zeta-function regularized determinants "det" instead of the heat-kernel determinant "Det", since with respect to integration on Teichmüller space the choice is irrelevant. Only when integrating with respect to the conformal parameter ψ (introduced in 6.9) does this choice have an influence on the result, as we pointed out in the remark at the end of section II.6.

We saw in section II.6 (cf. (6.9), (I.6.11)) that the renormalized Polyakov

118   II.7:   The Polyakov measure in the critical dimension $d = 26$

measure reads, after identifying $\det< \psi_g^i, \psi_g^j >_g$ with $\det< q_i, q_j >_g$:

$$d\mu_{\text{ren}}[\psi, t] = \frac{1}{Z_{\text{ren}}} \left(\frac{\text{Det}'(-\Delta_g)}{\int_S \sqrt{\det g}\, dz}\right)^{-d/2} \left(\frac{\text{Det}(P_g^* P_g)}{\det < q_i, q_j >_g}\right)^{\frac{1}{2}}$$
$$dt_1 \wedge \ldots \wedge dt_{3q-3} d[\psi] \qquad (7.1)$$
$$= \frac{1}{Z_{\text{ren}}} \left(\frac{\text{Det}'(-\Delta_g)}{\int_S \sqrt{\det g}\, dz}\right)^{-d/2} (\text{Det}(P_g^* P_g))^{\frac{1}{2}}\, d(WP) d[\psi].$$

Here S is the underlying differentiable surface, g is the metric determined by an element of $T_{p,k}$ and a conformal factor, $d[\psi]$ is the formal Lebesgue measure on the space of conformal factors, and $d(WP)$ is the Weil–Petersson measure on $T_{p,k}$.

When considering $T_{p,k}$ as a totally real subspace of T_q, we can identify $d(WP)$ with $\det< q_i, q_j >_{g_t}^{-\frac{1}{2}} dt_1 \wedge \ldots \wedge dt_{3q-3}$. Since the volume of $T_{p,k}$ with respect to $d(WP)$ is infinite, one expects the result of integrating with respect to the Teichmüller parameters over T_p to be infinite likewise.

In what follows we shall concentrate on the integration in the Teichmüller parameters and integrate only over M_p, which has finite volume, instead of T_p. (Note that the Weil–Petersson metric on T_p is invariant under the action of the mapping class group $\Gamma_p = D/D_0$; the singularities of M_p have to be handled by passing to a finite cover, as mentioned above in I.8).

The problem that arises here is the one of anomalies. Integrating a functional with respect to the Polyakov measure should yield a number in order to be physically meaningful as a probability density, but in the integrand we have sections of line bundles (because of the square root, even for the definition of $\det'\Delta_g$ we cannot just use the determinant of Δ_g which is a real number and have to use the determinant of the appropriate $\bar\partial$ operator instead, and similarly for $\text{Det}' P_g^* P_g$). When we quotient out the action of conformal transformations and of Γ_p, the resulting line bundle need not be trivial. First of all there may be a topological obstruction, namely a nonvanishing first Chern class. Following the discussion in [F1], this topological obstruction is called the global anomaly. Even if the line bundle is trivial, the trivialization need not be canonical – this, however, would be necessary in order to interpret sections as functions, namely as multiples of a fixed canonical section. A trivialization can only be chosen canonically if the bundle is flat. The obstruction to flatness is measured by the connection and curvature of a canonically defined metric, in our case the Quillen metric, and this obstruction is called the local or geometric anomaly. We now want to show that these anomalies vanish for $d = 26$.

As a preliminary remark, we note that we can choose holomorphic differentials on a given Riemann surface independently of a conformal factor for the metric – we shall always do so in the sequel.

II.7: *The Polyakov measure in the critical dimension $d = 26$*

We now write down the Quillen metrics of our determinant line bundles. On

$$\det P_g = \det \bar{\partial}_1 = L_2^* = \Lambda^{\max} (H^0(\Sigma, \tilde{\Omega}^2))^* \quad \text{(assuming } q' \geq 2\text{)}$$

a section is given by

$$s = (q_1 \wedge \ldots \wedge q_{3q-3})^{-1}, \tag{7.2}$$

and its Quillen metric is

$$\|s\|_Q^2 = \frac{\det' \bar{\partial}_1^* \bar{\partial}_1}{\det(q_i, q_j)}. \tag{7.3}$$

Likewise, a section of

$$\det \Delta = \Lambda^{\max} (H^0(\Sigma, \tilde{\Omega}))^* \otimes i\mathbb{R}$$

is given by

$$t = (\omega_1 \wedge \ldots \wedge \omega_q)^{-1} \otimes 1, \tag{7.4}$$

where $\omega_1, \ldots, \omega_q$ form a basis of the holomorphic 1-forms real on the boundary, and 1 denotes the constant section. Then

$$\|t\|_Q^2 = \frac{\det' \Delta}{\det(\omega_i, \omega_j) \int_\Sigma \sqrt{\det g}}. \tag{7.5}$$

We therefore rewrite (7.1) as

$$d\mu_{\text{ren}}[\psi, t] = \frac{1}{Z_{\text{ren}}} \|t\|_Q^{-d} \|s\|_Q^2 \det(\omega_i, \omega_j)^{-d/2} dt_1 \wedge \ldots \wedge dt_{3q-3} d\psi. \tag{7.6}$$

For technical reasons, we now turn again to the Schottky double Σ' of genus q'. We let g_0 be a metric on Σ' with curvature θ and $\lambda : \Sigma' \to \mathbb{R}$, and put

$$g_\lambda := e^{2\lambda} g_0.$$

We let $L_{n+1}^* = L_{n+1}^*(\Sigma')$ be the determinant line bundle associated to $\bar{\partial}_n$, and we let $\|\cdot\|_{Q,0}$ and $\|\cdot\|_{Q,\lambda}$ be the Quillen metrics on L_n defined through g_0 and g_λ, respectively We then have the following result of Bost and Jolicoeur [BJ]:

$$\frac{\|\cdot\|_{Q,\lambda}^2}{\|\cdot\|_{Q,0}^2} = \exp\left\{ \frac{6n(n-1)+1}{6\pi i} \int_\Sigma' (\lambda\theta + \partial\lambda \wedge \bar{\partial}\lambda) \right\}. \tag{7.7}$$

For the proof (sketched from [F2]), one considers the trivial holomorphic fibration

$$\pi : \mathbb{C} \times \Sigma' \to \mathbb{C},$$

chooses a real function $\Lambda(|z|,\cdot)$ on $\mathbb{C} \times \Sigma'$ with $\Lambda(0,\cdot) \equiv 1$, $\Lambda(1,\cdot) = 2\lambda$, and endows the fiber over z with the metric $e^{\Lambda(|z|,\cdot)}g_0$. In general, this is not a Kähler fibration, but it turns out that the canonical connection is compatible with the complex structure, and that (I.8.23) also holds for such a fibration. We thus let θ_π be the relative curvature of π, i.e. the curvature of the tangent bundle of the fibers, namely

$$\theta_\pi = \theta + \bar{\partial}\partial \Lambda. \tag{7.8}$$

We then have for the curvature of the Quillen metric

$$\begin{aligned}\frac{1}{2\pi i}\theta_Q &= c_1(L_{n+1}^*, \|\cdot\|_Q) \\ &= \int_{\text{fibers}} \text{ch}(-n\theta_\pi)\text{td}(\theta_\pi)\end{aligned} \tag{7.9}$$

where one again integrates the 4-form part.
From (7.8) and (7.9) one obtains

$$\begin{aligned}\theta_Q &= \frac{6n(n-1)+1}{24\pi i} \int_\Sigma^{'} (2\bar{\partial}\partial\Lambda \wedge \theta + \bar{\partial}\partial\Lambda \wedge \bar{\partial}\partial\Lambda) \\ &= \bar{\partial}_z\partial_z \left(\frac{6n(n-1)+1}{24\pi i}\int_\Sigma^{'} 2\Lambda_z\theta + \partial\Lambda_z \wedge \bar{\partial}\Lambda_z\right),\end{aligned} \tag{7.10}$$

with $\Lambda_z := \Lambda(z,.)$. Since $\theta_Q(z) = \bar{\partial}_z\partial_z \log \|\cdot\|_Q^2(z)$, and since $\|\cdot\|_Q(z)$ depends only on z, (7.7) follows.

The important conclusion from (7.7) is that the Quillen metric on

$$L_2^* \otimes L_1^{13}$$

is independent of the conformal factor $e^{2\lambda}$. Therefore, for dimension $d = 26$, the conformal anomaly disappears, and we consequently discard the integration over the conformal factors and redefine the Polyakov measure as (for $d = 26$)

$$d\mu[t] = \frac{1}{Z} \|t\|_Q^{-26} \|s\|_Q^2 \det(\omega_i, \omega_j)^{-13} dt_1 \wedge \ldots \wedge dt_{3q-3} \tag{7.11}$$

Now over T_q,

$$s \otimes t^{-13} = q_1 \wedge \ldots \wedge q_{3q-3} \otimes (\omega_1 \wedge \ldots \wedge \omega_q)^{-13}$$

is a global holomorphic section of the flat bundle

$$L_2^* \otimes L_1^{13},$$

II.7: The Polyakov measure in the critical dimension $d = 26$

and therefore it can naturally be interpreted as a holomorphic function on M_q. Restricting to $M_{p,k}$, the moduli space of surfaces of genus p with k boundaries, we obtain

$$\|t\|_Q^{-26} \, \|s\|_Q^2 = |f|^2,$$

where f is the restriction to $M_{p,k}$ of a holomorphic function on M_q. This has the consequence that for $d = 26$, all anomalies cancel. The line bundle $L_2^* \otimes L_1^{13}$ is topologically trivial on moduli space so that we can integrate over $M_{p,k}$ instead of $T_{p,k}$, and since the bundle is even flat, the section can be interpreted as a holomorphic function, so that we are now integrating a function over a space of finite volume.

Taking this into account, we redefine the Polyakov measure once more and put as our final answer for the case $d = 26$

$$d\mu[t] := |f|^2 \det(\omega_i, \omega_j)^{-13} \, dt_1 \wedge \ldots \wedge dt_{3q-3}, \tag{7.12}$$

which gives a precise meaning to the formal measure $h(t)dt$ arising in (6.11).

II.8 Correlation functions

As in section I.1, let Γ_1 and Γ_2 be two smooth closed oriented pointwise disjoint Jordan curves and let us consider the heuristic quantity called "amplitude" (related to the probability for the string to travel from Γ_1 to Γ_2):

$$Z(\Gamma_1,\Gamma_2) = \frac{1}{Z} \int_{\mathcal{F}\times M} e^{-D(X,g)} e^{-\mu^2 \int_S \sqrt{\det g}\, dz} dg[X]d[g] \qquad (8.1)$$

with the notation of section II.2 and where S is here a smooth compact surface with boundaries Γ_1, Γ_2 and $\Gamma_3,\ldots,\Gamma_k$, $k \geq 2$, the Γ_i being also smooth oriented pairwise disjoint curves. An embedding $X : S \to \mathbb{R}^d$ with the specified boundary conditions is split into the sum

$$X_0 + X,$$

where X_0 satisfies the boundary conditions and is harmonic with respect to the metric g, while X has boundary values zero (see section I.2). Thus, formally

$$Z(\Gamma_1,\Gamma_2) = \frac{1}{Z} \int e^{-D(X_0,g)} e^{-D(X,g)} d[X_0]d[X]d[g]$$

$$= \frac{1}{Z} \int e^{-D(X_0,g)} (\det{}'\Delta_g)^{-\frac{d}{2}} \left(\int \sqrt{\det g}\, dz\right)^{\frac{d}{2}} d[X_0]d[g] ,$$

after formally computing the Gaussian integral in the variable X as in section II.4. In order to evaluate the integration of $e^{-D(X_0,g)}$ we proceed as follows.

For each fixed conformal structure Σ on the surface S, and for any given conformal metric g on Σ, we choose an oriented diffeomorphism

$$\eta_g : (S^1)^k \to \partial\Sigma , \text{ where } (S^1)^k \text{ is the disjoint union of } k \text{ copies of } S^1,$$

which induces a metric g_η on $(S^1)^k$ in such a way that each γ_i is parametrized proportionally to arclength. Thus g_η is determined up to the action of the diffeomorphism group and a choice of scale. The ambiguity in the choice of diffeomorphism will leave our functional below invariant. The ambiguity in the choice of scale has to be treated differently, as the possible scaling factors vary in the noncompact space $(\mathbb{R}^+)^k$. We therefore fix length scales on $(S^1)^k$ by normalizing the metric g_n by the requirement that the length of the ith component of $(S^1)^k$ is the same as the length of the curve in the image configuration $\Gamma_1 \cup \Gamma_2$ corresponding to γ_i under our prescribed boundary conditions. Since the metric in the target space $\mathbb{R}^d$ yields the only fixed length scale available, this seems to be the only reasonable normalization; nevertheless, in our opinion, it is somewhat arbitrary (cf. [A] on this point).

II.8: Correlation functions

We now fix a conformal metric g_0 on Σ; this determines $\eta = \eta_{g_0}$ and the metric

$$g_\eta := \rho^2(t)dt^2$$

on $(S^1)^k$ by our length normalization. This induces a scalar product for tangent vector fields V, W of $(S^1)^k$, namely

$$(V,W) := \int_{(S^1)^k} \rho^2(t) \frac{dV}{dt} \frac{dW}{dt} \, dt.$$

This in turn induces a measure on $(\mathrm{Diff}_+ S^1)^k$, the group of oriented diffeomorphisms of $(S^1)^k$ fixing each component. We then define the formal Gaussian integral

$$j(\Gamma_1, \Gamma_2) := \int e^{-D(X_0 \circ \phi, g_0)} d[\phi] \tag{8.2}$$

where $d[\phi]$ is a formal Lebesgue measure on $(\mathrm{Diff}_+ S^1)^k$.

If one wants to avoid the difficulty that there may not be any nontrivial measure supported just on the diffeomorphism of S^1 one may also replace this integration by integration over the larger space of all orientation-preserving monotone selfmaps of $(S^1)^k$ fixing the components. (This is in analogy with the Wiener integral). We remark that $D(X_0 \circ \phi, g_0)$ is the Dirichlet integral of the harmonic extension with respect to g_0 of $X_{0|\partial\Sigma} \circ \phi$.

We want to check that $j(\Gamma_1, \Gamma_2)$ is independent of the choices of X_0 and g_0. It is independent of the choice of X_0, because any other boundary map is of the form $X_0 \circ \phi$, $\phi \in (\mathrm{Diff}_+ S^1)^k$, and we are integrating precisely over all such ϕ.

Independence on the choice of g_0 can be seen as follows. For given boundary values $\psi : \partial\Sigma \to \Gamma_1 \cup \Gamma_2$, we let

$$H(\psi, g)$$

be the harmonic extension of ψ with respect to g into Σ. We denote as before the Dirichlet integral of the harmonic extension of ψ with respect to g by $D(\psi, g)$. Then

$$D(X_0 \circ \phi, g_0) = \frac{1}{2} \int g_0^{ij}(z) \frac{\partial H(X_0 \circ \phi, g_0)(z)}{\partial z^i} \frac{\partial H(X_0 \circ \phi, g_0)(z)}{\partial z^j}$$
$$(\det g(z))^{\frac{1}{2}} dz$$
$$= \frac{1}{2} \int ((\phi^{-1})^* g_0)^{ij}(w) \frac{\partial H(X_0, (\phi^{-1})^* g_0)(w)}{\partial w^i}$$
$$\frac{\partial H(X_0, (\phi^{-1})^* g_0)(w)}{\partial w^j} (\det (\phi^{-1})^* g_0(w))^{\frac{1}{2}} dw$$

by a similar argument as in section I.3. Hence we infer that changing the metric by a diffeomorphism amounts to changing the boundary values by the inverse diffeomorphism (restricted to the boundary).

If we multiply g_0 by a conformal factor $\lambda^2(z)$, then this again leads to a change of η_{g_0} by a diffeomorphism of the boundary, while the product (V,W) above remains invariant, because it is invariant under diffeomorphisms by the same argument as in I.3, and the Dirichlet integral $D(X_0, g_0)$ is invariant under conformal changes of the metric g_0 as explained in I.1. Moreover the length scale is normalized independently of the conformal factor $\lambda^2(z)$. Finally, the ambiguity in the choice of initial points again is accounted for by the action of $(\text{Diff}_+ S^1)^k$.

In conclusion, $j(\Gamma_1, \Gamma_2)$ indeed is independent of all choices involved, except for the normalization of the length scales. $j(\Gamma_1, \Gamma_2)$ depends on the conformal structure Σ, however, hence on the variable t in Teichmüller space $T_{p,k}$. For this reason, we sometimes write $j_t(\Gamma_1, \Gamma_2)$.

For the evaluation of $j(\Gamma_1, \Gamma_2)$, it is useful to have an explicit formula for $D(Y, g_0)$ in terms of its boundary values on ∂S. If S is the unit disk D, then such a formula was already exhibited by J. Douglas, namely, with

$$\eta := Y_{|\partial D},$$

$$D(Y, g_0) = \frac{1}{16\pi} \int_0^{2\pi} \int_0^{2\pi} \frac{(\eta(\phi) - \eta(\phi'))^2}{\sin^2(\frac{\phi - \phi'}{2})} d\phi d\phi'$$

(cf. [JS]). Note that the metric g_0 does not enter into this formula, because of conformal invariance of the Dirichlet integral.

Since the remaining terms in the correlation functions are invariant under the action of the diffeomorphism group, we can implement the Faddeev–Popov procedure to redefine the correlation functions as heuristic path integrals:

$$Z(\Gamma_1, \Gamma_2) = \frac{1}{Z} \int_\varphi \int_{M_{p,k}} j(\Gamma_1, \Gamma_2) (\frac{\det \Delta'_g}{\sqrt{g}})^{-\frac{d}{2}} (\det P_g^* P_g)^{\frac{1}{2}} h(t) \, dt d[\varphi]$$

where $T_{p,k}$ is the Teichmüller space for surfaces of genus g with k boundary curves, $h(t) \, dt$ is a measure on $M_{p,k}$ and $j(\Gamma_1, \Gamma_2)$ is the t- but not φ-dependent term defined by (8.2). In the same way as we introduced the renormalization term $\exp(-\mu^2 \int_S \sqrt{g} dz)$, we can introduce an additional renormalization term $e^{-\lambda_0^2 \int_{\partial S} \sqrt{\tilde{g}} ds}$ in the expression (8.1) of the amplitude, where ∂S is the boundary of the surface S and $\tilde{g}$ is the induced metric on ∂S. The Laplace operator is taken here with Dirichlet boundary conditions. We obtain the "renormalized" amplitude:

$$\tilde{Z}(\Gamma_1, \Gamma_2) = \frac{1}{\tilde{Z}} \int_{\mathcal{F} \times M} e^{-D(X,g)} e^{-\mu^2 \int_S \sqrt{\det g} dz} e^{-\lambda_0^2 \int_{\partial S} \sqrt{\det \tilde{g}} dz} dg[X]d[g].$$

(8.3)

II.8: Correlation functions

We now let λ_0^2 tend in a suitable way to infinity so that in the limit $\lambda_0^2 \to \infty$ the term proportional to λ_0^2 compensates the divergent terms in the heat-kernel expansions arising from the boundary conditions. Proceeding as in the case of boundaryless surfaces, one can describe an ε-cutoff version of the "renormalized" amplitude $\tilde{Z}[\Gamma_1, \Gamma_2]$ as the formal integral $\int j_t(\Gamma_1, \Gamma_2) \mu_\varepsilon^b[\varphi, t]$ for an ε-cutoff heuristic measure defined by

$$d\mu_\varepsilon^b[\varphi, t] = \frac{Z_\varepsilon}{Z_\varepsilon^b} d\mu_\varepsilon[\varphi, t] \cdot e^{-\lambda_\varepsilon^2 \int_{\partial S} \sqrt{\det \bar{g}_t}\, dz}$$

where $\lambda_\varepsilon \in \mathbb{R}\backslash\{0\}$, $d\mu_\varepsilon[\varphi, t]$ is given by (6.1), $\bar{g}_t$ is the restriction of $\bar{g}_t$ to the boundary and $Z_\varepsilon^b = \int d\mu_\varepsilon[\varphi, t] e^{-\lambda_\varepsilon^2 \int_{\partial S} \sqrt{\det \bar{g}_t}\, dz}$.

Setting as in (6.6) $\lambda_\varepsilon^2 = \frac{a}{\sqrt{\varepsilon}} + b\lambda_0$, a and b well chosen and defined in terms of the dimension d, and using Propositions I.11.2 and I.11.3 extended to the case of Dirichlet boundaries, we can interpret the "renormalized" amplitude $\tilde{Z}(\Gamma_1, \Gamma_2)$ as

$$\frac{1}{Z} \int j_t(\Gamma_1, \Gamma_2) e^{-\mathcal{L}_b(\psi, t)} h(t) d[\psi] dt$$

where $h(t)$ was defined in (6.8) and $h(t)dt$ can be interpreted as $d\mu[t]$ as in (7.12) and where $e^{-\mathcal{L}_b(\psi, t)} d[\psi]$ is a formal measure describing a Liouville model with boundary terms which we identify with the measure $\nu_\alpha^{c,b}$ described in section I.10.

As our final answer we obtain, for $d = 26$, $\tilde{Z}(\Gamma_1, \Gamma_2) = \int_{M_{p,k}} j_t(\Gamma_1, \Gamma_2) d\mu[t]$, where μ is the measure on $M_{p,k}$ given by (7.12)

References

[A]   O. Alvarez, Theory of strings with boundaries: Fluctuations, topology, and quantum geometry. Nucl. Phys. B 216, 125–184 (1983)

[AB]   S. Albeverio, Z. Brzeźniak, Finite dimensional approximations approach to oscillatory integrals and stationary phase in infinite dimensions. J. Funct. Anal. 113, 177–244 (1993)

[AFHKL] S. Albeverio, J.E. Fenstad, R. Høegh-Krohn, T. Lindstrøm, Nonstandard methods in stochastic analysis and mathematical physics, Pure and Applied Mathematics 122, Academic Press, New York (1986)

[AGGSIS] L. Alvarez-Gaumé, M.B. Green, M.T. Grisaru, R. Iengo, E. Sezgin, Edts., Superstrings '87, World Scientific, Singapore ('87)

[AHK]   S. Albeverio, R. Høegh-Krohn, Mathematical Theory of Feynman Path Integrals. Lect. Notes Math. 523, Springer, Berlin (1976)

[AHK1]   S. Albeverio, R. Høegh-Krohn, The Wightman axioms and the mass gap for strong interactions of exponential type in two dimensional space-time. J. Funct. Anal. 16, 39–82 (1974)

[AHK2]   S. Albeverio, R. Høegh-Krohn, Uniqueness and the global Markov property for Euclidean fields. The case of trigonometric interactions, Comm. Math. Phys. 70, 187–192 (1979)

[AHK3]   S. Albeverio, R. Høegh-Krohn, Diffusion fields, quantum fields, and fields with values in Lie groups, pp. 1–98 in "Stochastic Analysis and Applications", Eds. M.A. Pinsky, M. Dekker, New York (1984)

[AHKH]   S. Albeverio, R. Høegh-Krohn, H. Holden, Markov cosurfaces and gauge fields, Acta Phys. Austr. Suppl. XXVI, 211–231 (1984)

[AHKHK] S. Albeverio, R. Høegh-Krohn, H. Holden, T. Kolsrud, Representation and construction of multiplicative noise, J. Funct. Anal. 87, 250–272 (1989)

[AHKMTT] S. Albeverio, R. Høegh-Krohn, J. Marion, D. Testard, B. Torresani, Non commutative distribution theory, M. Dekker (1992)

[AHKPS1] S. Albeverio, R. Høegh-Krohn, S. Paycha and S. Scarlatti, Path space measure for the Liouville quantum field theory and the construction of the relativistic strings. Phys. Lett. B174, 81–86 (1986)

[AHKPS2] S. Albeverio, R. Høegh-Krohn, S. Paycha, S. Scarlatti, A probability measure for random surfaces of arbitrary genus and bosonic strings in 4 dimensions, in Proceedings "Eugene Wigner Symposium on Space Time Symmetries" Washington 1988, Proceedings Series Nuclear Phys. B6, 180–182 (1989)

[AHKPS3] S. Albeverio, R. Høegh-Krohn, S. Paycha, S. Scarlatti, A global and stochastic analysis approach to bosonic strings and associated quantum fields, Acta Applic. Math. 26, 103–195, (1992)
[AHKZ] S. Albeverio, R. Høegh-Krohn, B. Zegarliński, Uniqueness and global Markov property for Euclidean fields: The case of general polynomial interactions, Commun. Math. Phys. 123, 377–424 (1989)
[AIK] S. Albeverio, K. Iwata, T. Kolsrud, Random fields as solutions of the inhomogeneous quaternionic Cauchy-Riemann equations. I. Invariance and analytic continuation, Comm.Math.Phys. 132, 555–580 (1990)
[AN] L. Alvarez-Gaumé and P. Nelson, Riemann Surfaces and String Theories, pp. 419–510 in "Supersymmetry, supergravity and superstrings ('86 Trieste)", Eds. B. DeWitt, M. Grisaru, Singapore World Scientific 1986
[AP1] M. Arnaudon, S. Paycha, Factorisation of semi-martingales on infinite dimensional principal bundles, to appear in Stochastics and Stochastic Reports (1995)
[AP2] M. Arnaudon, S. Paycha, Regularisable and minimal orbits for group actions in infinite dimensions and projections of regularised Brownian motion, Strasbourg preprint (1995)
[APS] S. Albeverio, S. Paycha, S. Scarlatti, A short overview of mathematical approaches to functional integration, pp. 230–276 in "Functional integration, geometry and strings", Eds. Z. Haba, J. Sobczyk, Birkhäuser (1989)
[AS] S. Albeverio, J. Schäfer, Abelian Chern–Simons theory and linking numbers via oscillatory integrals, in "Special Issue on Functional Integration" J. Math. Phys. 36 2157–2169 (1995)
[B] L. Bers, Riemann surfaces, Lectures, Courant Institute of Mathematical Sciences, New York (1957-59)
[BDVH] L. Brink, P. Di Vecchia, P. Howe, A locally supersymmetric and reparametrization invariant action for the spinning string, Phys. Lett. 65 B, 471–474, (1976)
[Be] C. Becker, J. Funct. Anal. 134, 321–349 (1995)
[BeGV] N. Berline, E. Getzler, M. Vergne, Heat kernels and Dirac operators, 2nd ed., Springer-Verlag Berlin (1996)
[BF] J. Bismut and D. Freed, The analysis of elliptic families, I, Comm. Math. Phys. 106, 159–176 (1986)
[Bill] P. Billingsley, Probability and measure. Wiley, New York, (1979)
[BJ] J.B. Bost and T. Jolicoeur, A holomorphic property and the critical dimension in string theory from an index theorem, Phys. Lett. B174, 273–276 (1986)
[BK] A. Belavin, V.G. Knizhnik, Algebraic geometry and the geometry of quantum strings. Phys. Lett. B168, 201–206 (1986)

[BM] A.A. Beilinson, Yu.I. Manin, The Mumford form and the Polyakov measure in string theory. Comm. Math. Phys. 107, 359–376 (1986)

[BrH] L. Brink, M. Henneaux, Principles of string theory, Plenum Press, New York (1988)

[Ca] J.L. Cardy, Conformal invariance and statistical mechanics, Lecture Notes in Les Houches, Session XLIX (1988)

[Cha] I. Chavel, Eigenvalues in Riemannian Geometry. Academic Press, London (1984)

[D] B. Durhuus, Quantum theory of strings. Nordita lectures (1982)

[DdD] G.F. De Angelis, D. de Falco and G. Di Genova, Random fields on Riemannian manifolds: A constructive approach, Comm. Math. Phys. 103, 297–303 (1986)

[DHP1] E. D'Hoker and D.H. Phong, Multiloop amplitudes for the bosonic Polyakov string. Nucl. Phys. B269, 205–234 (1986)

[DHP2] E. D'Hoker and D.H. Phong, The geometry of string perturbation theory Rev. Modern Phys. 60, 917–1065 (1988)

[DNOP] B. Durhuus, H.B. Nielsen, P. Olesen, J.L. Peterson, Dual models as saddle point approximations to Polyakov's quantized string, Nucl. Phys. B196, 498–508 (1982)

[DOP] B. Durhuus, P. Olesen and J.L. Petersen, Polyakov's quantized string with boundary terms I. Nucl. Phys. B198, 157–188 (1982); II: B201, 176–188 (1982)

[Eb] D. Ebin, The manifold of Riemannian metrics. Proc. Symp. Pure Math. AMS 15, 11–40 (1970)

[EE] C.J. Earle and J. Eells, A fibre bundle description of Teichmüller theory. J. Diff. Geom. 3, 19–43 (1969)

[F] D. Friedan, Introduction to Polyakov's string theory, pp. 839–867 in "Recent advances in field theory and statistical mechanics", Eds. R. Stora and J.B. Zuber, Elsevier (1984)

[F1] D. Freed, Determinants, torsion, and strings, Comm. Math. Phys. 107, 483–513 (1986)

[F2] D. Freed, On determinant line bundles, in: S.T. Tau (ed.), Mathematical aspects of string theory, pp. 189–238, World Scientific, Singapore (1987)

[FK] H.M. Farkas and I. Kra, Riemann Surfaces, Springer-Verlag Berlin (1980)

[FP] L.D. Faddeev, V.N. Popov, Feynman Diagrams for the Yang-Mills fields, Phys. Lett. B25, 29–30 (1967)

[G1] P. Gilkey, The Index Theorem and the Heat Equation, Mathem. Lect. Ser. No.4, Publish or Perish, Boston, MA (1974)

[G2] P. Gilkey, Invariance Theory, the Heat Equation and the Atiyah-Singer Index Theorem, Mathem. Lect. Ser., 11, Publish or Perish (1984)

[GGRT] P. Goddard, J. Goldstone, C. Rebbi, C.B. Thorn, Quantum dynamics of a massless relativistic string, Nucl. Phys. 56, 109–135 (1973)

[GH] P. Griffiths and J. Harris, Principles of algebraic geometry, Wiley, New York (1978)

[Gi] G. Gilbert, String theory path integral: Genus two and higher. Nucl. Phys. B277, 102–124 (1986)

[GJ] J. Glimm, A. Jaffe, Quantum Physics, A Functional Integral Point of View, 2nd Edition, Springer-Verlag, New York (1987)

[Go] T. Goto, Relativistic quantum mechanics of one-dimensional mechanical continuum and subsidiary condition of dual resonance model, Progr. Theor. Phys. 46, 1560–1569 (1971)

[Gro] L. Gross, Harmonic analysis on Hilbert space. Memoirs Am. Math. Soc. no. 46 AMS, Providence RI (1963)

[GrKS] L. Gross, C. King, A. Sengupta Two-dimensional Yang–Mills theory via stochastic differential equation, Ann Phys. 164, 65–112 (1989)

[Gr] D.J. Gross, The status and future prospects of string theory, Nucl. Phys. B (Proc. Suppl.) 15, 43–56 (1990)

[GrSW] M.B. Green, J.H. Schwarz, E. Witten, Superstring Theory, Cambridge University Press, Cambridge (1987)

[GV] I.M. Gelfand, N.Ya. Vilenkin, Generalized Functions Vol 4. N.Y. Academic Press 1964

[H] R. Hartshorne, Algebraic Geometry, Springer-Verlag Berlin, Heidelberg, New York (1977)

[Hab] Z. Haba, Stochastic equations for gauge fields, Journ. of Phys. A18, n.15, L957-L962 (1985)

[Hida] T. Hida, Brownian motion, Springer-Verlag, Berlin (1980)

[HK] R. Høegh-Krohn, A general class of quantum fields without cut-off in two dimensional space-time. Comm. Math. Phys. 21, 244–255 (1971)

[I] K. Ito, Generalized uniform complex measures in the Hilbertian metric space with their application to the Feynman path integral, Proc. Vth Berkeley Symp. on Mathematical Statistics and Probability, Univ. of California Press, Berkeley (1967), Vol. II, pp. 55–161

[ID] C. Itzykson, J.M.Drouffe, Théorie Statistique des Champs I,II Editions du C. N. R. S., Savoirs actuels (1989)

[J1] J. Jost, Harmonic maps between surfaces, Springer Lect. Notes Math. 1062, (1984)

[J2] J. Jost, Two-dimensional geometric variational problems, Wiley–Interscience, Chichester (1991)

[J3] J. Jost, Strings with boundary: a quantization of Plateau's problem, SFB 237 Preprint (1988)

[J4] J. Jost, Harmonic mappings between Riemannian manifolds, ANU Press, Canberra (1984)

[Ja] R. Jackiw, Liouville field theory: a two dimensional model for gravity, in Quantum theory of gravity, Ed. S. Christensen, Adam Hilger, Bristol, 403–420 (1983)

[Jas1] Z. Jaskólski, On the Gribov ambiguity in the Polyakov string. J. Math. Phys. 29, 1034–1034 (1988)

[Jas2] Z. Jaskólski, The integration of G-invariant functions and the geometry of the Faddeev–Popov procedure. Comm. Math. Phys. 111, 439–468 (1987)

[Jas3] Z. Jaskólski, The Polyakov path integral over bordered surfaces (the open string amplitudes), Comm. Math. Phys. 128, 285–318 (1990)

[Jas4] Z. Jaskólski, Liouville gravity on bordered surfaces, Int. J. Mod. Phys. A8, 1041–1057 (1993)

[JM] Z. Jaskólski, K. Meissner, Static quark potential from the Polyakov sum over surfaces, Nucl. Phys. B 418, 456–476 (1994)
Z. Jaskolski, K. Meissner, First quantized noncritical relativistic Polyakov string, Nucl. Phys. B 428, 331–373 (1994)

[JS] J. Jost and M. Struwe, Morse–Conley theory for minimal surfaces of varying topological type, Invent. Math. 102, 465–499 (1990)

[K] T.P. Killingback, Global aspects of fixing the gauge in the Polyakov string and Einstein gravity. Comm. Math. Phys. 100, 267–277 (1985)

[Ka] V. Kac, Infinite dimensional Lie algebras, Birkhäuser, Cambridge, MA (1983)

[Kah] J.P. Kahane, Sur le chaos multiplicatif, Ann. Sci. Math. Québec 9(2), 105–150 (1985)

[Kn] V.G. Knizhnik, Multiloop amplitudes in the theory of quantum strings and complex geometry, Sov. Sci. Rev. A Phys. 10, 1–76 (1989) (transl.)

[Ku] S. Kusuoka, Høegh-Krohn's model of quantum fields and the absolute continuity of measures, pp. 405–424 in "Ideas and Methods in Quantum and Statistical Physics", R. Høegh-Krohn Memorial Vol., Eds. S. Albeverio, J.E. Fenstad, H. Holden, T. Lindstrøm, Cambridge Univ. Press (1992)

[Kuo] H. H. Kuo, Gaussian measures in Banach Spaces. Lect. Not. Mat. 463, Berlin, Springer (1975)

[L] O. Lehto, Univalent Functions and Teichmüller Spaces, Springer-Verlag, Berlin (1987)

[Le] J. Lepowski, Ed., Algebras, lattices and strings, Springer-Verlag Berlin (1984)

[M] J. Milnor, Remarks on Infinite–Dimensional Lie Groups. pp.

References

1007–1057 in "Relativity, Groups and Topology II" Les Houches Session XL 1983, Eds. B.S. DeWitt and R. Stora, Elvesier (1984)

[Ma1] Yu.I.H. Manin, Quantum strings and algebraic curves, ICM Berkeley Lecture pp. 1286–1295, Proc ICM Berkeley, AMS, Providence (1986)

[Ma2] Yu.I.H. Manin, Strings, Math. Intell. 11, 59-65 (1989)

[Mi] J. Mickelsson, Current algebras and groups, Plenum Press, New York (1989)

[MN] G. Moore and P. Nelson, Measure for Moduli. Nucl. Phys. B266, 58–74 (1986) (Uspekhi Math. Nauk 30:1 (1975), pp. 3–59)

[MoP] A.Yu. Morozov, A.M. Perelomov, String theory and complex geometry, Phys. Reports (1992)

[N] P. Nelson, Lectures on Strings and Moduli Space. Phys. Reports 149, 337–375 (1987)

[Ne] E. Nelson, Feynman integrals and the Schrödinger equation, J. Math. Phys. 5, 332–343 (1964)

[O] H. Omori, On the group of diffeomorphisms on a compact manifold. Proc. Symp. Pure Math. AMS 15, 167–183 (1970)

[P1] A.M. Polyakov, Quantum geometry of bosonic strings. Phys. Lett. B103, 207–210 (1981)

[P2] A.M. Polyakov, Quantum geometry of fermionic strings, Phys. Lett. B103, 211–213 (1981)

[P3] A.M. Polyakov, Gauge fields and strings, Harwood Ac. Publ., Contemporary Concepts in Physics (1987)

[Pa1] S. Paycha, A mathematical interpretation of the Polyakov string model in non critical dimensions, Thesis, Paris (1990)

[Pa2] S. Paycha, The Faddeev–Popov procedure and applications to bosonic strings: an infinite dimensional point of view, Comm. Math. Phys. 147, 163–180 (1992)

[Pa3] S. Paycha, Elliptic operators in the functional quantisation of gauge field theories, to appear in Comm. Math. Phys.

[Po1] J. Polchinski, Evaluation of one loop string path integral. Comm. Math. Phys. 104, 37–47 (1986)

[Po2] J. Polchinski, From Polyakov to Moduli. pp. 13–28 in "Mathematical Aspects of String Theory" Proceedings of San Diego Conference July 1986, Eds. Yau, Singapore World Scientific 1987 (Adv. Ser. Math. Phys. 1)

[PS] A. Presley, G. Segal, Loop groups, Clarendon Press, Oxford (1986)

[RaSi] D.B. Ray, I.M. Singer, R-torsion and the Laplacian on Riemannian manifolds. Adv. Math. 7, 145–210 (1971)

[S1] B. Simon, The $P(\phi)_2$ Euclidean Quantum Field Theory, Princeton University Press (1974)

[S2] B. Simon, Trace ideals and their applications, Cambridge University Press 1979 (London Math. Soc. Lect. Notes 35)

[S3] B. Simon, Schrödinger Semigroups, Bull AMS 7, 447–526 (1982)

[Sc] S. Scarlatti, Metodi analitici e probabilistici nello studio del modello di Polyakov per la stringa bosonica, Thesis, Roma (1989)

[Sch] A.S. Schwartz, Instantons and fermions in the Field of Instanton, Comm. Math. Phys. 64, 233–268 (1978/79)

[See] R. Seeley, Complex powers of an elliptic operator, in Proc. Symp. Pure Math. Vol 10, Am. Math. Soc. Providence (1967), 288–307

[Shu] M. Shubin, Pseudo-differential operators and spectral theory. Springer-Verlag (1987)

[Sko] A.V. Skorohod, Integration in Hilbert Spaces. Springer-Verlag (1974)

[Sm] D.J. Smit, String theory and algebraic geometry of moduli spaces, Comm. Math. Phys. 114, 645–685 (1988)

[T1] A.J. Tromba, Global Analysis and Teichmüller Theory. pp. 167–198 in "Seminar on new results in non linear partial differential equations" A.J. Tromba, Public. Max-Planck Inst. für Math. Bonn, Vieweg-Braunschweig (1987)

[T2] A.J. Tromba, Teichmüller theory in Riemannian geometry, Birkhäuser, Basel (1992)

[Ta] M.E. Taylor, Pseudodifferential operators, Princeton University Press (1981)

[W] S. Weinberg, Covariant Path Integral Approach to String Theory. Lectures given at the 3rd Jerusalem Winter School of Theoretical Physics 1987, UTTG-17-87

[Wa] R.M. Wald, On the Euclidean Approach to Quantum Field Theory in Curved Spacetime. Comm. Math. Phys. 70, 221–242 (1979)

[Wo] S. Wolf, The Teichmüller theory of harmonic maps, Thesis, Stanford (1986)

[Wp] S. Wolpert, Chern forms and the Riemann tensor for the moduli space of curves, Inv. Math. 85, 119–145 (1986)

[Yos] K. Yosida, Functional analysis. Springer-Verlag, Berlin (1978)

Index

A
Abstract Wiener space 70
Actional functional 1, 106
Affine Hilbert space 70
Almost complex structure 30
Amplitudes 115
Annulus 30
Anomalies 118
Area functional A 7

C
Cauchy-Riemann operators 31, 32, 57
Characteristic function (of a random vector) 60
Characteristic functional 67
Chern character 62
Chern classes 59, 61, 118
Chern polynomial 62
Chow ring 61
Classical string theory 1
Conformal anomaly 120
Conformal factors 120
Conformal fields 4
Conformal group 97
Conformal structure Σ 31
Conformal parameters 9
Conformal transformations 8
Consistency property 73
Correlation functions 100, 122, 117
Cosmological term, constant 105, 114
Countable Hilbert space 71
Covariance matrix 66
Covariance operator 67
Critical dimension 3
Curvature 120
Cylinder set, measure 68

D
$\mathcal{D}^l$ diffeomorphisms of Sobolev class H^l 12
$\mathcal{D} = \cap_{l \geq 2} \mathcal{D}^l$ 13
$\mathcal{D}_0^l$ Connected component of identity in $\mathcal{D}^l$ 21
$\mathcal{D}_0$ Connected component of identity in $\mathcal{D}$ 26

Δ_g Laplace-Beltrami operator 32
$\bar{\partial}$ 119
$D(X, g)$ 8, 106
$D(X, \Sigma)$ 9
Determinants 36, 48
Determinant bundle 48, 59, 119
Dirac operators 49
Distribution 66
Divisor 59

E
ε - regularized heat kernel determinant 39
ε - cut-off Polyakov measure 113, 114
Ell$^\rightarrow$ positive elliptic self-adjoint operators 39
Energy functional 21
Energy momentum tensor 8
Energy of a map 21
Euclidean quantum field theory 82
Euler-Lagrange equations 7, 24
Expectation 66
Exponential Chern character 62
Exponential interaction / Høegh-Krohn model 76

F
$\mathcal{F}^s$ 11
Faddeev-Popov determinant 41, 44, 45, 46
Faddeev-Popov map, procedure, operator 41, 43, 46, 98, 104, 109, 124
Feynman (path integrals) measure 92
Finite dimensional distributions 73
Formal Lebesgue measure 99, 101
Formal renormalized measure 116
Free Euclidean field 74
Free Euclidean field measure 70-74

G
g metric 8
Gauge 28
Gaussian distribution 66, 68, 69, 71
Gaussian generalized stochastic process 74
Gaussian measure 66

Gauss measure associated with a Hilbert space 68
Gaussian random vectors 67
Generalized random field 74
Generalized stochastic process 74
Global anomaly 118

H
Harmonic Beltrami differential 30
Harmonic gauge 29
Harmonic map 1, 8, 21
Heat-kernel determinant of an operator 40
Hodge bundle 64
Hodge divisor class 64
Høegh-Krohn model 3, 76
Holomorphic line bundle 59
Holomorphic quadratic differential 9, 16, 22, 118

I
ILH-Lie group 17
ILH-manifold 17
Isothermal coordinates 15, 24

K
Kac–Moody algebras 2

L
Lagrangian of Liouville model 115
Laplace–Beltrami operator Δ_g 32
Line bundle 60
Liouville measure (with constant curvature term and a boundary term) 83
Liouville model 83, 115
Local anomaly 118

M
$\mathcal{M}^u$ Riemannian metrics of class H^k 12
$\mathcal{M}$ smooth Riemannian metrics 13
$\mathcal{M}_q$ universal modular curve 63
Mapping class group 49, 63
Mean 66
Measurable norm 69
Minimal surface 8
Moduli space 63

N
$n(g)$ 98

Noncritical dimension 113
Nuclear space 72

P
P_g maps of vector fields into symmetric traceless 2×2 tensors 32
Picard group 60
Plateau problem 1, 4
Polyakov's approach 2
Polyakov measure 2, 96
Positive definite functional 67
Probability measure 67

Q
$Q(\Sigma)$ holomorphic quadratic differentials on Σ and on $\partial\Sigma$ 23
Quantization 2, 92
Quantum fields 2
Quillen metric 52, 65, 118, 119

R
Random field 66
Regularized Faddeev–Popov determinant 47
Relativistic string 1
Renormalized amplitude 125
Renormalized heat-kernel determinant 40
Riemann surface 31
Riemannian metric (space of) 12
Riemann moduli space M_q 63
Rigging of a Hilbert space 70

S
S (compact oriented two-dimensional differentiable manifold) 11
Scalar curvature 38
Schottky double 14, 21, 30, 35, 54, 119
Small time expansion 88
Sobolev space $H^{k,p}$ 25
Standard gaussian distribution 67

T
Teichmüller space 21, 63
Todd class 62

U
Universal modular curve 63
Universal Teichmüller curve 49

V
Version of gaussian measure associated with Hilbert space 70
Virasoro algebras 2

W
Weil–Petersson metric 17, 65
Weyl estimate 52

X
X embeddings 2, 7

Z
Zeta-function determinant of an operator 38
Zeta-function of an operator 36
Zeta-function regularization 52

Printed in the United States
By Bookmasters